Nomenclator

Alliorum

Nomenclator Alliorum

Allium names and synonyms – a world guide

COMPILED BY

Mary Gregory, Reinhard M. Fritsch, Nikolai W. Friesen, Furkat O. Khassanov and Dale W. McNeal

Production Editor: S. Dickerson

Cover design by Jeff Eden

Page make-up by Media Resources,
Information Services Department,
Royal Botanic Gardens, Kew

ISBN 1 900347 64 4

Printed by
Whitstable Litho Printers Ltd.,
Whitstable, Kent, U.K.

Contents

Introduction

Allium is the largest genus of petaloid monocotyledons, excluding orchids, with some 750 species according to Stearn (1992). There has been no comprehensive monograph of the genus since Regel's in 1875 and the taxonomy is complicated, with a proliferation of synonyms and disagreement as to the subdivision of the genus. The fact that important characters are often lost in herbarium specimens, so that a study of living material is essential, adds to the problem.

Allium includes economically important food crops such as onions, garlic, leeks and chives, also species with medicinal properties and others of horticultural merit. There is considerable interest in investigating wild relatives of these plants for plant breeding and possible future genetic manipulation. Workers studying any of these subjects need to know if the name they are using is the generally accepted one. These lists aim to fulfill this need and to serve as a working tool. They may also indicate to taxonomists where there are problems worthy of investigation and thus help to stimulate further research. The first list includes all except American species and the second list is of American species. About 1400 species names are included in the two lists, of which at least half are generally considered to be synonyms.

There is only one species, *Allium schoenoprasum*, common to both geographical areas, and therefore the Floras used for the two lists are completely different. This made it sensible to compile separate lists. The arrangement of the two lists is also somewhat different because the American species of subgenus *Amerallium* have not been further subdivided, as McNeal explains in his introductory remarks.

These lists give some indication of species distribution also. However, it must be remembered that some of the Floras were published decades ago and there have been many more recent records as well as several corrections.

List of *Allium* species names (excluding American species)

compiled by

Mary Gregory
Jodrell Laboratory, Royal Botanic Gardens, Kew, Richmond, Surrey TW9 3AB, UK,

Reinhard M. Fritsch and **Nikolai W. Friesen**
Institut für Pflanzengenetik und Kulturpflanzenforschung, Corrensstrasse 3, D-06466 Gatersleben, Germany

and **Furkat O. Khassanov**
Institute of Botany, Uzbek Academy of Sciences, Tashkent, Uzbekistan.

This list is an attempt to help people find out the generally accepted names of Old World *Allium* species and to which section a species of *Allium* belongs, by compiling a table showing the sections assigned to each species in six Floras that cover much of the area concerned, plus taxonomic revisions of certain sections, and additional records from *Index Kewensis* and a few other sources. It includes about 1170 species names, over half of which are generally considered to be synonyms.

Column 1 lists all species names found in the works examined (but not all subspecific taxa are included) with authorities. Sometimes there are slight differences in spellings or authorities; these have been checked against the *Index Kewensis* (IK) and sometimes in the original source, and the incorrect spelling or letters are enclosed in [ ], e.g. *achaium* [*-aicum*].

The next six columns show in which Floras the name appears, its species number and section in that work. The Floras are:

Column 2. *Flora Reipublicae Popularis Sinicae* (ed. F.T.Wang & T.Tang), vol. 14, 1980. Academia Sinica. *Allium* by J.M.Xu. (As listed in: *Key to the Alliums of China*, by J.M.Xu, P.Hanelt & C.-L.Long 1990. *Herbertia* **46**, 140-164).

Column 3. Conspectus of the wild growing *Allium* species of Middle Asia by F.O.Khassanov. Pp. 141-159 in: *Plant life in Southwest and Central Asia* (eds. M.Öztürk, Ö.Seçmen & G.Görk). Proc. IV Plant life of Southwest Asia Symposium, May 1995. Ege Univ. Press: Izmir vol. I, 1996 (publ. 1997).

Column 4. *Flora of the USSR*, vol. IV. Liliiflorae and Microspermae (ed. V.L.Komarov). Transl. from Russian by N.Landau 1968. Israel Program for Scientific Translations: Jerusalem. *Allium* by A.I.Vvedensky.

+ = additional or differing records from *Opredelitel' rastenii Srednei Azii [Conspectus Florae Asiae Mediae]* (ed. S.S.Kovalevskaya), vol. II, 1971. Izdat. FAN uzbek. SSR: Tashkent. *Allium* by A.I.Vvedensky.

Column 5. *Flora Iranica* (ed. K.H.Rechinger). Part 76, Alliaceae by P.Wendelbo 1971. Akademische Druck- und Verlagsanstalt: Graz.

Column 6. *Flora of Turkey*, vol. 8 (ed. P.H.Davis) 1984. Edinburgh Univ. Press. *Allium* by F.Kollmann; *Nectaroscordum* by Kit Tan. Some additions from supplementary vol. 10 (1988).

Column 7. *Flora Europaea* (ed. T.G.Tutin *et al.*), vol. V, 1980. Cambridge Univ. Press. *Allium* by W.T.Stearn.

Column 8 lists other records, mostly from B.E.E. de Wilde-Duyfjes 1976. A revision of the genus *Allium* L. (Liliaceae) in Africa. *Meded. Landbouwhog. Wageningen* 76 (11),

239 pp.; these are marked *. Also from Pal = *Flora Palaestina*, vol. 4 (ed. N.Feinbrun-Dothan). Israel Academy of Sciences: Jerusalem. *Allium* by F.Kollmann. Ohwi = *Flora of Japan*, by J.Ohwi 1984. Smithsonian Institution: Washington. 1067 pp. [Eng. edn.] (No sections listed in this work.) Other reports in this column have a brief reference in column 11; the full references are given on p. 68.

Column 9 gives the sections for spp. of subgenus *Melanocrommyum* as revised by R.M.Fritsch and of subgenus *Rhizirideum* as revised by N.W.Friesen (see also p. 67).

Column 10 lists the species numbers for section *Allium* species as given in B.Mathew's taxonomic review (Mathew, B. 1996. *A review of Allium section Allium*. Royal Botanic Gardens, Kew.)

Column 11 (last) gives the names for synonyms, as recognized in the various Floras. For example, *Fl. USSR* and *Fl. Europ.* both consider *angulosum* is the correct name for *acutangulum*, so it appears as =*angulosum*, but *Fl. Turk.* considers *frigidum* is the correct name for *achaium*, whereas *Wilde-Duyfjes regards *paniculatum* as correct, so it appears as Turk=*frigidum*, *=*paniculatum*. We are merely reporting these opinions, not suggesting that any of them is necessarily correct. However, names recognized by B.Mathew (columnn 10) or by RMF, NWF and FOK can be accepted as representing good modern taxonomic judgements.

Species names in *Index Kewensis* but not in any of these Floras, either because they are in an area not covered by the Floras, e.g. Burma, India, or because they were described after publication of the relevant Floras, have been listed with a note of the area of distribution and publication date as given in IK in column 11, and the section name if deducible from the original publication in column 8. A few older names, which are in Regel's Monograph (1875) or in *Index Kewensis* but are not included in the synonomy in any of the relevant Floras and are of doubtful status, have been omitted.

If readers encounter names not in the list, we should be pleased to receive details of the publications in which they were seen.

A second table lists the sections recognized in the various Floras, with the species numbers for each section. *Fl. USSR*, following Regel's monograph of 1875, puts many species in *Molium*, which would be in other sections in most Floras. The latest proposed subgeneric grouping by Hanelt *et al.* (1992) at Gatersleben divided the genus into six subgenera and about 45 sections. Prior to this most workers recognized between five and 30 sections of very different standing and often contradictory circumscription. Therefore, even when the section names in columns 2–7 of the first table are identical, the authors have not always treated the sections in the same way, and comments in column 11 refer only to the status of the taxa in the Floras indicated.

Acknowledgements

Brian Mathew (Kew) initially suggested which Floras to include and advised MG on correct spelling of species names, for which she is grateful. Jill Cowley (Kew), and Peter Hanelt (Gatersleben) have made valuable suggestions and Rosemary Davies (Kew) has helped answer some queries with regard to *Index Kewensis*. Thanks are due to Vivien Munday for typing the tables.

Allium species and synonyms (excluding American species)

Column 1 Species of Allium & authorities (incl. synonyms)	2 Fl. China	3 Middle Asia	4 Fl. USSR (+= Fl. As. med.)	5 Fl. Iran	6 Fl. Turkey	7 Fl. Europ.	8 Others	9 Fritsch & Friesen	10 B. Math-ew	11 Accepted names for synonyms
a.-bolosii Palau Ferrer	—	—	—	—	—	—	—	—	—	see antonii-bolosii
achaium Boiss. et Orph. [-aicum]	—	—	—	—	31 Scor	—	14 Cod*	—	—	Turk=frigidum,*=paniculatum
" var. parnassicum Boiss.	—	—	—	—	—	—	64 Cod	—	—	=parnassicum (Stearn 1978)
acidoides Stearn	—	—	—	—	—	—	Col	—	—	IK (Burma) 1960; RMF=Col
aciphyllum J.M.Xu	37 Rhiz	—	—	—	—	—	—	Ret	—	
acutangulum Cat. or Schrad.	—	—	37 Rhiz	—	—	1 Rhiz	—	Rhiz	—	=angulosum
acutiflorum Lois.	—	—	—	—	—	85 All	—	—	11	
adzharicum M.Pop.	—	—	—	—	—	—	Rhiz	Ret	—	IK (USSR) 1941 =pseudostrictum (Kud 1992)
aegaeum Heldr. et Halácsy ex Halácsy	—	—	—	—	—	—	4a All*	—	62a	*=sphaerocephalon ssp. sphaerocephalon; BM=?ssp. sphaer.
aemulans Pavl.	—	110 Scor	+ Hap	—	—	—	—	—	—	+=caesium
aestivale J.J.Rodr. [-is]	—	—	—	—	—	—	—	—	2	=commutatum
aethusanum Garbari	—	—	—	—	—	—	Scor	—	—	Miceli & Garbari 1991; Sicily
affine Boiss. et Heldr.	—	—	—	—	104 All	95 All	1 All*	—	85	=vineale (or v. var. virens)
" Ledeb.	—	—	156 Porr	70 All	107 All	—	—	—	82	
" var. scabrum (Regel) Grossh.	—	—	—	—	—	—	—	—	82	=affine
afghanicum Wend.	—	6 Ret	—	1 Rhiz	—	—	—	Ret	—	
aflatunense B.Fedtsch.	—	149 Meg	209 Mol	—	—	—	—	Meg	—	
" hort. non Fedtsch.	—	—	—	—	—	—	—	Meg	—	=hollandicum
africanum Dietrich	—	—	—	—	—	—	25 Mol*	—	—	=roseum
afrum Kunth	—	—	—	—	—	—	29b Mel*	Mel	—	RMF=nigrum; *=n. ssp. nigrum
aitchisonii Boiss.	54 Rhiz	45 Orei	—	11 Rhiz	—	—	—	Orei	—	=carolinianum
" Regel	—	113 Avu	—	44 Scor	—	—	—	—	—	=umbilicatum
akaka Gmel. ex Roem. et Schult.	—	—	190 Mol	98 Acan	120 Acan	—	—	Acan	—	
" ssp. haemanthoides (Boiss. et Reut. ex Regel) Wend.	—	—	—	—	—	—	Acan	Acan	—	=haemanthoides; IK 1973
" " shelkovnikovii (Grossh.) Wend.	—	—	—	—	—	—	Acan	Acan	—	=shelkovnikovii; IK 1973

alabasicum (D.S.Wen et S.Chen) Y.Z.Zhao	—	—	—	—	—	—	?Rhiz	Rhiz	—	IK (Inner Mongolia) 1992
alaicum Vved.	—	155 Acmo	206 Mol	—	—	—	—	Acmo	—	
alaschanicum Y.Z.Zhao	—	—	—	—	—	—	?Rhiz	Rhiz	—	IK (China) 1992
alataviense Regel	53 Rhiz	—	54 Rhiz	—	—	—	—	Orei	—	=platyspathum (NWF= " ssp. amblyophyllum)
albanum Grossh.	—	112 Avu	115 Hap	43 Scor	—	43 Scor	—	—	—	=rubellum (Asia, Iran, Eur)
albertii Regel	88 Hap	121 Caer	127 Hap	—	—	—	—	—	—	=pallasii
albescens Guss.	—	—	—	—	—	—	—	—	1	=ampeloprasum
albidum Fisch. ex Bess.	—	—	38 Rhiz	—	—	6b Rhiz	—	—	—	Eur=albidum ssp. caucasicum
" " " Bieb.	—	—	—	—	—	6 Rhiz	—	Rhiz	—	
" ssp. albidum	—	—	—	—	—	6a Rhiz	—	Rhiz	—	
" " caucasicum (Regel) Stearn	—	—	—	—	3 Rhiz	6b Rhiz	—	Rhiz	—	
" Ledeb.	—	—	39 Rhiz	—	—	—	—	—	—	=flavescens
albiflorum Omelczuk	—	—	—	—	—	88 All	—	—	33	
albopilosum Wright [albipilosum]	—	192 Kal	192 Mol + Mol	102 Acan	121 Acan	—	—	Kal	—	=cristophii [christophii] (+=bodeanum)
albostellerianum Wang et Tang	40 Rhiz	—	—	—	—	—	—	Sacc	—	China=paepalanthoides
albotunicatum O. Schwarz	—	—	—	—	67 Cod	—	Pal 14 Cod	—	—	
" ssp. albotunicatum	—	—	—	—	67 Cod	—	Pal 14 Cod	—	—	
albovianum Vved.	—	—	76 Rhiz	—	—	—	—	Orei	—	
album Santi	—	—	—	—	12 Mol	28 Mol	23 Mol*	—	—	=neapolitanum
" var. purpurascens Maire et Weiller	—	—	—	—	—	—	18b Mol*	—	—	*=subhirsutum ssp. subvillosum (=subvillosum in Stearn 1978)
" auct. non Santi	—	—	—	—	—	—	18b Mol*	—	—	*=subhirsutum ssp. subvillosum
alexandrae Vved.	—	64 Orei	73 Rhiz	—	—	—	—	Orei	—	
alexeianum Regel [alexejanum, alexjanum]	—	195 Kal	189 Mol	—	—	—	—	Kal	—	see Kh & Fr 1994
" var. hissaricum Lipsky	—	—	—	93 Acan				Kal		=nevskianum
alibile A.Rich.	—	—	—	—	—	—	7 All*	—	31	*=ampeloprasum
alpinarii Özhatay et Kollm.	—	—	—	—	28 Scor	—	—	—	—	
alpinum (DC.) Heget.	—	—	—	—	—	18 Schoe	—	Schoe	—	=schoenoprasum var. alpinum
altaicum Pall.	75 Schoe	82 Cep	87 Phyll	—	—	—	—	Cep	—	
altissimum Regel	—	—	210 Mol	117 Meg	—	—	—	Meg	—	see also Fritsch 1993, 1996
altyncolicum Friesen	—	—	—	—	—	—	Schoe	Schoe	—	IK (Altai, USSR) 1987
amabile Stapf	42 Rhiz	—	—	—	—	—	Col	—	—	RMF=Col
amamianum Tawada	—	—	—	—	—	—	?	?	—	IK (Ryukyu Is.) 1975
ambiguum Sm.	—	—	—	—	15 Mol	23 Mol	25 Mol*	—	—	=roseum

Column 1 Species of Allium & authorities (incl. synonyms)	2 Fl. China	3 Middle Asia	4 Fl. USSR (+= Fl. As. med.)	5 Fl. Iran	6 Fl. Turkey	7 Fl. Europ.	8 Others	9 Fritsch & Friesen	10 B. Mathew	11 Accepted names for synonyms
amblyanthum Zahar.	—	—	—	—	41 Cod	57a Cod	Pal 10 Cod	—	—	=pallens ssp. pallens
amblyophyllum Kar. et Kir.	53 Rhiz	44 Orei	54 Rhiz	—	—	—	—	Orei	—	=platyspathum (Asia= " ssp. amblyophyllum)
amethystinum Tausch	—	—	—	—	105 All	96 All	2 All*	—	81	*=guttatum
" var. antiquorum Radić	—	—	—	—	—	—	All	—	—	Radić 1989
ammophilum Heuffel	—	—	—	—	—	6a Rhiz	—	Rhiz	—	Eur=albidum ssp. albidum NWF=flavescens
amoenum G.Don	—	—	—	—	—	23 Mol	—	—	—	=roseum
ampeloprasoides Miscz. ex Grossh.	—	—	—	—	92 All	—	—	—	25	=gramineum
" Grossh.	—	—	164 Porr	—	—	—	—	—	—	=fominianum
ampeloprasum L.	—	85 All	175 Porr	82 All	76 All	76 All	7 All*, Pal 15 All	—	1	see also Stearn 1978
" ssp. bimetrale (Gand.) Hayek	—	—	—	—	79 All	81 All	7 All*	—	2	=commutatum (*=ampeloprasum)
" " commutatum (Guss.) Zangheri	—	—	—	—	—	—	—	—	2	=commutatum
" " eu-ampeloprasum Markgr. or Hayek	—	—	—	—	—	—	7 All*	—	1	=ampeloprasum
" " iranicum Wend.	—	—	—	82b All	—	—	—	—	5	BM=iranicum
" " leucanthum (C.Koch) Hayek	—	—	—	—	—	76 All	—	—	6	Eur perhaps=leucanthum or commutatum; BM=leucanthum
" " polyanthum (Schult. et Schult.f.) O. de Bolos et al.	—	—	—	—	—	—	—	—	10	=polyanthum
" " porrum (L.) Hayek	—	—	—	—	—	76 All	—	—	—	Eur=ampeloprasum; NWF=porrum
" " truncatum (Feinbr.) Kollm.	—	—	—	—	—	—	Pal 16 All	—	18	=truncatum
" var. atroviolaceum (Boiss.) Regel	—	84 All	173 Porr	—	83 All	—	—	—	13	=atroviolaceum
" " babingtonii (Borrer) Syme	—	—	—	—	—	76 All	—	—	—	
" " bulbiferum Syme	—	—	—	—	—	76 All	—	—	—	
" " bulbilliferum Lloyd	—	—	—	—	—	76 All	—	—	—	=ampeloprasum var. bulbiferum
" " caucasicum	—	—	173 Porr	—	—	—	—	—	—	=atroviolaceum
" " caudatum Pamp.	—	—	—	—	—	—	7 All*	—	1	*=ampeloprasum
" " combazianum Maire	—	—	—	—	—	—	7 All*	—	1	*=ampeloprasum
" " commutatum (Guss.) Fiori	—	—	—	—	79 All	81 All	—	—	2	=commutatum

" " duriaeanum (Gay) Bonnet et Barratte	—	—	—	—	—	—	9 All*	—	48	=baeticum
" " gracilis Cavara	—	—	—	—	—	—	7 All*	—	1	*=ampeloprasum
" " holmense Asch. et Graeb.	—	—	—	—	—	76 All	7 All*	—	1	*=ampeloprasum
" " leucanthum Regel or (C.Koch) Ledeb.	—	—	174 Porr	—	76 All	—	7 All*	—	6	USSR & BM=leucanthum Turk & *=ampeloprasum
" " lussinense Haračić	—	—	—	—	79 All	—	—	—	2	=commutatum
" " porrum (L.) J.Gay	—	—	176 Porr	—	77 All	76 All	7 All*, Pal s.n. All	—	1a	=porrum (*=ampeloprasum)
" " " Regel	93 Porr	—	—	—	—	—	—	—	—	=porrum
" " portorii Gomb.	—	—	—	—	—	—	Pal 16 All	—	18	=truncatum
" " pruinosum Boiss.	—	—	—	—	79 All	—	7 All*	—	2	=commutatum (*=ampeloprasum)
" " truncatum Feinbr.	—	—	—	—	—	—	—	—	18	=truncatum
" " wiedemannii Regel	—	—	—	—	76 All	—	7 All*	—	1	=ampeloprasum
" Waldst. et Kit. (non L.)	—	—	—	—	89 All	—	—	—	—	=scorodoprasum ssp. waldsteinii
amphibolum Ledeb.	—	—	16 Rhiz	—	—	—	—	Ret	—	
" auct. non Ledeb. p.p.	—	—	—	—	—	—	Rhiz	—	—	=gubanovii (Friesen 1988)
" " " " "	—	—	—	—	—	—	Rhiz	—	—	=malyschevii " "
amphipulchellum Zahar.	—	—	—	—	—	74 Cod	—	—	—	perhaps=stamineum
anacoleum Hand.-Mazz.	—	—	—	28 Scor	25 Scor	—	—	—	—	
anatolicum Özhatay et B.Mathew	—	—	—	—	—	—	—	—	97	SW.Turkey
andersonii G.Don	51 Rhiz	—	44 Rhiz	—	—	—	—	Rhiz	—	China & USSR=senescens, NWF=angulosum
angolense Baker	—	—	—	—	—	—	10 Schoe*	Cep	—	=cepa
anguinum Bubani	—	—	—	—	—	—	22 Ang	—	—	=victorialis (Stearn 1978)
angularum var. caucasicum Regel	—	—	38 Rhiz	—	—	—	—	—	—	=albidum
angulosum L.	—	37 Rhiz	37 Rhiz	—	—	1 Rhiz	—	Rhiz	—	
" ssp. ammophilum (Heuffel) K.Richter	—	—	—	—	—	—	6a Rhiz	—	—	=albidum ssp. albidum (Stearn 1978)
" var. calcareum Wallroth	—	—	—	—	—	—	3b Rhiz	—	—	=senescens ssp. montanum (Stearn 1978)
" " caucasicum Regel	—	—	—	—	3 Rhiz	6b Rhiz	—	—	—	=albidum ssp. caucasicum
" " minum Ledeb. [-nus]	51 Rhiz	—	44 Rhiz	—	—	—	—	—	—	=senescens
" auct. non L.	—	—	—	—	—	—	Rhiz	—	—	=dauricum (Friesen 1988)
angustitepalum Wend.	—	146 Meg	—	118 Meg	—	—	—	Meg	—	Iran=rosenbachianum, Asia & RMF =jesdianum ssp. angustitepalum
angustum G. Don	—	—	—	—	—	—	—	Ret	—	NWF=strictum

Column 1 Species of Allium & authorities (incl. synonyms)	2 Fl. China	3 Middle Asia	4 Fl. USSR (+= Fl. As. med.)	5 Fl. Iran	6 Fl. Turkey	7 Fl. Europ.	8 Others	9 Fritsch & Friesen	10 B. Mathew	11 Accepted names for synonyms
anisopetalum Vved.	—	—	118 Hap	—	—	—	—	—	—	=anisotepalum [sphalm.]
anisopodium Ledeb.	47 Rhiz	43 Ten	51 Rhiz	—	—	—	—	Ten	—	
" ssp. argunense Peschkova	—	—	—	—	—	—	—	Ten	—	IK (Siberia) 1979; NWF=anisopodium
" var. zimmermannianum (Gilg) Wang et Tang	47a Rhiz	—	—	—	—	—	—	—	—	
anisotepalum Vved. [Wend.]	—	130 Minu	+ Hap	48 Scor	—	—	—	—	—	+ & Iran=griffithianum
antiatlanticum Emberger et Maire	—	—	—	—	—	—	14 Cod*	—	—	=paniculatum
antonianii Bordz.	—	—	—	—	—	—	—	Camp	—	NWF=scabriscapum
antonii-bolosii Palau Ferrer	—	—	—	—	—	—	49b Scor	—	—	=cupani ssp. hirtovaginatum (Stearn 1978); see also Garbari et al. 1991
" " ssp. eivissanum (Miceli et Garbari) N.Torres et Rosselló	—	—	—	—	—	—	?	—	—	=eivissanum; IK (Balearic Is.) 1988
anzalonei Brullo, Pavone et Salmeri	—	—	—	—	—	—	Cod	—	—	IK (Italy) 1997
aobanum Araki	—	—	—	—	—	—	Cep	Cep	—	IK (Japan) 1955; NWF=cepa
araxanum Fom. ex Grossh.	—	—	98 Hap	—	21 Brev	—	—	—	—	USSR= lacerum; Turk=callidictyon Not in IK
arenarium L.	—	—	—	—	—	—	—	—	85	=vineale
argyi Lévl.	25 Rhiz	—	—	—	—	—	—	But	—	=tuberosum
aristatum Cand.	—	—	—	—	s.n. Cod	—	—	—	—	imperfectly known
arlgirdense Blakelock	—	—	—	29 Scor	27 Scor	—	—	—	—	
armenum Boiss. et Kotschy	—	—	—	—	62 Cod	—	—	—	—	
armerioides Boiss. [armeroides]	—	—	—	—	114 All	—	—	—	90	
aroides M.Pop. et Vved.	—	181 Aroi	186 Mol	—	—	—	—	Aroi	—	Kh & Fr 1994
artemisietorum Eig et Feinbr.	—	—	—	—	—	—	5 All* Pal 21 All	—	91	*=ascalonicum
artvinense Miscz. ex Grossh.	—	—	—	—	103 All	—	—	—	74	
arvense Guss.	—	—	—	—	95 All	90b All	4a All*	—	62c	=sphaerocephalon ssp. arvense (*= " ssp. sphaerocephalon)
" var. trachypus (Boiss. et Spruner) Halácsy	—	—	—	—	95 All	—	—	—	62b	=sphaerocephalon ssp. trachypus
" auct.	—	—	—	—	—	—	4c All*	—	—	= " ssp. curtum

ascalonicum L.	78 Cep	—	—	—	s.n. Schoe	—	5 All* Pal 20 All	Cep	88	BM & Pal =hierochuntinum, NWF=cepa var. aggregatum
" auct. (non L.)	—	—	—	—	—	20 Cep	10 Schoe*	Cep	—	=cepa
ascendens Ten.	—	—	—	—	—	—	7 All*	—	1	=ampeloprasum
aschersonianum W.Barbey	—	—	—	—	134 Mel	—	30 Mel* Pal 25 Mel	Mel	—	*=orientale
" ssp. ambiguum Bég. et Vaccari	—	—	—	—	—	—	30 Mel*	Mel	—	*=orientale, RMF=aschersonianum
" " tel-avivense (Eig) Opphr.	—	—	—	—	—	—	30 Mel* Pal 26 Mel	Mel	—	Pal & RMF=tel-avivense, *=orientale
" var. caudatum Täckh. et Drar	—	—	—	—	—	—	7 All*	—	—	=ampeloprasum
asclepiadeum Bornm.	—	—	—	—	133 Mel	—	—	Mel	—	
" auct. (non Bornm.)	—	—	—	—	—	—	Pal 24 Mel	—	—	=nigrum
asirense B.Mathew	—	—	—	—	—	—	—	—	16	Saudi Arabia (BM)
asperiflorum Miscz. ex Grossh.	—	—	—	—	90 All	—	—	—	42	
assimile Halácsy	—	—	—	—	104 All	95 All	—	—	85	=vineale; Eur=v. var. virens
atriphoeniceum Bornm.	—	—	—	—	128 Mel	—	—	Pseud	—	=cardiostemon
atropurpureum Waldst. et Kit.	—	—	—	—	123 Mel	105 Mel	29a Mel*	Mel	—	* =nigrum ssp. multibulbosum
" var. hirtulum Regel	—	—	—	120 Meg	140 Mel	—	—	Meg	—	Iran & Turk=hirtifolium, RMF=stipitatum
atrosanguineum Kar. et Kir. or Schrenk	71 Schoe	70 Ann	81 Rhiz	—	—	—	—	Ann	—	USSR=monadelphum; NWF=good sp.
atroviolaceum Boiss. [atriv-]	—	84 All	173 Porr	81 All	83 All	79 All	—	—	13	
" var. caucasicum Somm. et Lev.	—	—	173 Porr	—	—	—	—	—	13	=atroviolaceum
" " firmotunicatum [(Fom.)] Grossh.	—	—	—	—	—	79 All	—	—	13	= "
" " ruderale Grossh.	—	—	—	—	—	—	—	—	13	= "
aucheri Boiss.	—	—	154 Porr	66 All	109 All	—	—	—	58	
auctum Omelczuk	—	—	—	—	—	106 Mel	—	Mel	—	RMF=cyrilli/decipiens; Eur may =nigrum
aureum Lam.	—	—	—	—	—	—	33 Mol	—	—	=moly (Stearn 1978)
auriculatum Kunth	—	—	—	—	—	—	17 Rhiz	Ret	—	Hook. rev. Stearn 1945
austrosibiricum Friesen	—	—	—	—	—	—	Rhiz	Rhiz	—	IK (Tuva; Mongolia) 1987
autumnale P.H.Davis	—	—	—	—	—	—	Cod	—	—	IK (Cyprus) 1949
azaurenum Gomb.	—	—	—	—	—	—	Pal 12 Cod	—	—	perhaps=sindjarense
azureum Ledeb.	90 Hap	117 Caer	129 Hap	—	—	—	—	—	—	=caeruleum
" var. bulbiferum Schrenk [-bulbilif-]	—	117 Caer	+ Hap	—	—	—	—	—	—	=caeruleum
babingtonii Borrer	—	—	—	—	—	76 All	—	—	—	=ampeloprasum var. babingtonii

Column 1 Species of Allium & authorities (incl. synonyms)	2 Fl. China	3 Middle Asia	4 Fl. USSR (+= Fl. As. med.)	5 Fl. Iran	6 Fl. Turkey	7 Fl. Europ.	8 Others	9 Fritsch & Friesen	10 B. Mathew	11 Accepted names for synonyms
backhousianum Regel	—	152 Acmo	—	—	—	—	Mol	Acmo	—	IK (Himalayas) 1885; Kh & Fr 1994
badakhshanicum Wend.	—	—	—	124 Meg	—	—	—	Brevic	—	RMF=pauli
baeticum Boiss.	—	—	—	—	—	84 All	9 All*	—	48	
bahri Regel	—	—	121 Hap	—	—	—	—	—	—	=griffithianum
baicalense Willd.	51 Rhiz	—	44 Rhiz	—	—	—	—	Rhiz	—	=senescens
baissunense Lipsky	—	191 Kal	220 Mol + Mol	—	—	—	—	Kal	—	USSR=caspium; Asia & RMF=c. ssp. baissunense; see Kh & Fr 1994
bajtulinii M.S.Baitenov et I.I.Kamenetskaya	—	54 Orei	—	—	—	—	Rhiz	Orei	—	IK (USSR) 1990; Asia =kastekii
bakeri Regel	81 Hap	—	—	—	—	—	—	Sacc	—	=chinense
" var. morrisonense (Hayata) T.S.Liu et S.S.Ying	—	—	—	—	—	—	?	—	—	=morrisonense; IK (Taiwan) 1978
bakhtiaricum Regel [bacht-]	—	—	—	?	—	—	—	Meg	—	Iran=incomplete specimen; see Fritsch 1996
balansae Boiss.	—	—	—	—	32 Scor	—	—	—	—	
baluchistanicum Wend.	—	—	—	3 Rhiz	—	—	—	Ret	—	
barsczewskii Lipsky [barszcz-]	—	26 Camp	27 Rhiz	8 Rhiz	—	—	—	Camp	—	
barthianum Asch. et Schweinf.	—	—	—	—	—	—	5 All*	—	93	*=ascalonicum
baschkyzylsaicum Krassovsk.	—	173 Acmo	—	—	—	—	—	Acmo	—	=taeniopetalum ssp. mogoltavicum (Fritsch et al. 1998)
bassitense Thiéb.	—	—	—	—	42 Cod	—	—	—	—	
bauerianum Baker	—	—	—	—	—	—	29a Mel*	Mel	—	*=nigrum ssp. multibulbosum RMF=nigrum
baytopiorum Kollm. et Özhatay	—	—	—	—	117 All	—	—	—	104	
beckerianum Regel	—	—	32 Rhiz	—	—	—	—	Camp	—	=inderiense
beesianum W.W. Smith	33 Rhiz	—	—	—	—	—	—	Ret	—	
bellulum Prokh.	—	—	49 Rhiz	—	—	—	—	Caes	—	
bidentatum Fisch. ex Prokh.	31 Rhiz	41 Caes	48 Rhiz	—	—	—	—	Caes	—	
biflorum Nakai	—	—	—	—	—	—	Mol	—	—	IK (Korea, Manchuria) 1913
bimetrale Gand.	—	—	—	—	79 All	81 All	7 All*	—	2	= commutatum (*=ampeloprasum)
birkinshawii Mouterde	—	—	—	—	—	—	?Cod	—	—	IK (Syria) 1970
blandum Wall.	54 Rhiz	45 Orei	+ Rhiz	11 Rhiz	—	—	—	Orei	—	=carolinianum
blomfieldianum Asch. et Schweinf.	—	—	—	—	—	—	27 Mol*	—	—	

bodeanum Regel	—	—	193 Mol	103 Acan	—	—	—	Kal	—	Iran perhaps=cristophii
bogdoicolum Regel	23 Rhiz	3 Ret	15 Rhiz	—	—	—	—	Ret	—	China=strictum
boissieri Regel	—	101 Mult	+ Porr	59 All	—	—	—	—	100	=borszczowii (Asia= good sp.)
bolosii Palau	—	—	—	—	—	49b All	—	—	—	See antonii-bolosii
borbasii A.Kerner	—	—	—	—	—	—	90a All	—	—	=sphaerocephalon ssp. sph. (Stearn 1978)
bornmuelleri Hayek	—	—	—	—	—	42 Scor	—	—	—	
borszczowii Regel	—	100 Mult	145 Porr	59 All	—	—	—	—	100	
boryanum Kunth	—	—	—	—	—	—	50 Scor	—	—	=callimischon ssp. callimischon (Stearn 1978)
bosniacum Kummer et Sendtner [-nai-]	—	—	—	—	—	—	90a All	—	—	=sphaerocephalon ssp. sph. (Stearn 1978)
botschantzevii R.Kamelin	—	151 Meg	—	—	—	—	?Mel	Meg	—	IK (Uzbekistan) 1976
bouddhae O.Debeaux	—	—	—	—	—	—	—	—	—	NWF=?fistulosum (not in Fl.China)
bourgeaui Rech.f.	—	—	—	—	78 All	80 All	—	—	3	
" ssp. bourgeaui	—	—	—	—	78 All	80a All	—	—	3a	
" " creticum Bothmer	—	—	—	—	—	80c All	—	—	3c	
" " cycladicum Bothmer	—	—	—	—	78 All	80b All	—	—	3b	
brachyodon Boiss.	—	7 Ret	18 Rhiz	2 Rhiz	—	—	—	Ret	—	
brachyscapum Vved.	—	199 Acan	194 Mol	114 Meg	—	—	—	Acan	—	
brachystemon Redouté [-um]	—	—	—	—	—	—	18a Mol*	—	—	=subhirsutum ssp. subhirsutum
bracteolatum Wend.	—	—	—	37 Scor	—	—	—	—	—	
brahuicum Boiss.	—	191 Kal	—	131 Kal	—	—	—	Kal	—	=caspium
brevicaule Boiss. et Bal.	—	—	—	—	51 Cod	—	—	—	—	
brevicuspis Boiss.	—	—	—	—	—	—	—	—	44	=erubescens
brevidens Vved.	—	95 Brevid	146 Porr	—	—	—	—	Brevid	110	see Fritsch et al. 1998
brevidentatum F.Z.Li	—	—	—	—	—	—	?Rhiz	Ret	—	IK (China) 1986
brevidentiforme Vved.	—	96 Brevid	+ Porr	—	—	—	—	Brevid	111	see Fritsch et al. 1998
brevipes Ledeb.	—	—	154 Porr	—	109 All	—	—	—	58	=aucheri
breviradium (Halácsy) Stearn	—	—	—	—	—	27 Mol	—	—	—	
breviscapum Stapf	—	—	—	96 Acan	—	—	—	Acan	—	
bucharicum Regel [bunch-]	—	194 Kal	219 Mol	130 Kal	—	—	—	Kal	—	USSR=schubertii Vved.; RMF=good sp.
buhseanum Regel	—	—	—	17 Schoe	5 Schoe	—	—	Schoe	—	=schoenoprasum
bulgaricum (Janka) Prodan.	—	—	—	—	Nect	Nect	—	—	—	=Nectaroscordum siculum ssp. bulgaricum
bulleyanum Diels	10 Brom	—	—	—	—	—	—	—	—	=wallichii
" var. tchongchanense (Levl.) Airy-Shaw	—	—	—	—	—	—	Rhiz	—	—	Airy-Shaw 1931

Column 1 Species of Allium & authorities (incl. synonyms)	2 Fl. China	3 Middle Asia	4 Fl. USSR (+= Fl. As. med.)	5 Fl. Iran	6 Fl. Turkey	7 Fl. Europ.	8 Others	9 Fritsch & Friesen	10 B. Math-ew	11 Accepted names for synonyms
bunchariсum Regel										see bucharicum
bungei Boiss.	—	—	—	34 Scor	—	—	—	—	—	
burjaticum Friesen	—	—	—	—	—	—	Rhiz	Rhiz	—	IK (S.Siberia; Mongolia) 1987
byzantinum C.Koch	—	—	—	—	—	—	7 All*	—	1	*=ampeloprasum, BM=?ampel.
cabulicum Baker	—	—	—	cult.	—	—	—	Mini	—	RMF=karataviense?
caerulescens Boiss. [coer-]	—	—	—	—	109 All	—	—	—	58	=aucheri
" G.Don [coer-]	90 Hap	—	129 Hap	—	—	—	—	—	—	=caeruleum [coer-]
caeruleum Pall. [coer-]	90 Hap	117 Caer	129 Hap	—	—	54 Scor	—	—	—	
caesioides Wend.	—	—	—	24 Scor	—	—	—	—	—	
caesium Schrenk	—	118 Caer	130 Hap	23 Scor	—	—	—	—	—	
" auct.	—	—	+ Hap	—	—	—	—	—	—	=litvinovii
caespitosum Siev. ex Bong. et Mey.	30 Rhiz	—	52 Rhiz	—	—	—	—	—	—	
" Siev. ex Pall.	—	40 Rhizo	—	—	—	—	—	Rhizo	—	
calabrum (N.Terracc.) Brullo, Pavone et Salmeri	—	—	—	—	—	—	Cod	—	—	IK (Italy) 1994
calamarophilon D.Phitos et Tzanoud.	—	—	—	—	—	—	Scor	—	—	IK (Greece) 1981
calcareum Reuter	—	—	—	—	—	—	—	—	—	RMF=?carinatum
callidictyon C.A.Mey. ex Kunth	—	—	97 Hap	21 Scor	21 Brev	—	—	—	—	
" auct.	—	—	+ Porr	—	—	—	—	—	—	=borszczowii
callimischon Link	—	—	—	—	24 Brev	50 Scor	—	—	—	
" ssp. callimischon	—	—	—	—	—	50a Scor	—	—	—	
" " haemostictum Stearn	—	—	—	—	24 Brev	50b Scor	—	—	—	
callistemon Webb ex Regel	—	—	—	—	55 Cod	—	—	—	—	= flavum ssp. flavum var. minus
calocephalum Wend.	—	—	—	110 Mel	—	—	—	Acmo	—	
calyptratum Boiss.	—	—	—	—	91 All	—	—	—	45	
candargyi Karavokyrou et Tzanoud.	—	—	—	—	—	—	Cod	—	—	IK (Greece) 1994
candidissimum Cavanilles	—	—	—	—	—	—	23 Mol*	—	—	=neapolitanum
candolleanum Albov	—	—	179 Mol	—	13 Mol	—	—	—	—	Turk perhaps=zebdanense
cannaefolium Lévl.	6 Ang	—	—	—	—	—	—	—	—	=prattii
capillare Cav.	—	—	—	—	30 Scor	—	—	—	—	=moschatum
capitellatum Boiss.	—	—	—	27 Scor	—	—	—	—	—	

cappadocicum Boiss.	—	—	—	—	84 All	—	—	—	15	
caput-medusae Airy-Shaw	—	—	—	—	—	—	Ang	Ang	—	IK (Burma) 1931
cardiostemon Fisch. et Mey.	—	—	197 Mol	112 Mel	128 Mel	—	—	Pseud	—	
carduchorum C.Koch	—	—	—	—	109 All	—	—	—	58	=aucheri
caricifolium Kar. et Kir.	88 Hap	121 Caer	127 Hap	—	—	—	—	—	—	=pallasii
caricoides Regel	—	—	65 Rhiz	—	—	—	—	Orei	—	
" "	—	53 Orei	+ Rhiz	—	—	—	—	—	—	=kokanicum
carinatum L.	—	—	103 Hap	—	58 Cod	71 Cod	—	—	—	
" ssp. carinatum	—	—	—	—	58 Cod	71a Cod	—	—	—	
" " pulchellum (G.Don) Bonnier et Layens	—	—	—	—	58 Cod	71b Cod	—	—	—	
" var. capsuliferum Ledeb.	—	—	102 Hap	—	—	—	—	—	—	=carinatum ssp. pulchellum (Stearn 1978)
" " " (Koch) Koch	—	—	—	—	—	—	71b Cod	—	—	= " " " " "
" " pulchellum Fiori	—	—	—	—	—	—	71b Cod	—	—	= " " " " "
carmeli Boiss.	—	—	—	—	—	—	Pal 8 Mol	—	—	
carneum Targ.-Tozz.	—	—	—	—	15 Mol	23 Mol	25 Mol*	—	—	=roseum
caroli-henrici Wend.	—	—	—	134 Thaum	—	—	—	Thaum	—	
carolinianum DC. [ex Redouté]	54 Rhiz	45 Orei	—	11 Rhiz	—	—	—	Orei	—	DC. is correct; IK has [Delar.]
caspium (Pall.) M.Bieb.	—	191 Kal	220 Mol	131 Kal	—	110 Kal	—	Kal	—	
" ssp. baissunense (Lipsky) Khassanov et R.M.Fritsch	—	191 Kal	—	—	—	—	—	Kal	—	Kh & Fr 1994
" ssp. caspium	—	191 Kal	—	—	—	—	—	Kal	—	Kh & Fr 1994
cassium Boiss.	—	—	—	—	11 Mol	—	—	—	—	
" var. hirtellum Boiss.	—	—	—	—	11 Mol	—	—	—	—	=cassium
cathodicarpum Wend.	—	—	—	138 Reg	—	—	—	Reg	—	
caucasicum M.Bieb.	—	—	71 Rhiz	—	—	—	—	Orei	—	=saxatile
" Ker-Gawl.	—	—	72 Rhiz	—	—	—	—	—	—	=globosum
cepa L. (cult.)	77 Cep	—	93 Cep	s.n. Cep	7 Cep	20 Cep	10 Schoe* Pal s.n. Cep	Cep		
" var. proliferum Regel	77a Cep	—	—	—	—	—	—	Cep	—	=A. × proliferum (Moench.) Schrad. (pers. comm. P.Hanelt)
" " " (Moench) Targ. Tozz.	—	—	—	—	—	—	20 Cep	Cep	—	=cepa (Stearn 1978)
" " sylvestre Regel	—	80 Cep	92 Cep	—	—	—	—	Cep	—	=oschaninii
" " " auct.	—	—	+ Phyll	—	—	—	—	—	—	=praemixtum or vavilovii
cepaeforme G.Don	—	—	—	—	—	—	20 Cep	Cep	—	=cepa (Stearn 1978)
chalcophengos Airy-Shaw	71 Schoe	—	—	—	—	—	—	Ann	—	=atrosanguineum

Column 1 Species of Allium & authorities (incl. synonyms)	2 Fl. China	3 Middle Asia	4 Fl. USSR (+= Fl. As. med.)	5 Fl. Iran	6 Fl. Turkey	7 Fl. Europ.	8 Others	9 Fritsch & Friesen	10 B. Math-ew	11 Accepted names for synonyms
chalkii Tzan. et Kollm.	—	—	—	—	—	—	Scor	—	—	IK (Aegean Is.) 1991
chamaemoly auct.	—	—	—	—	—	—	18b Mol*	—	—	=subhirsutum ssp. subvillosum
" L.	—	—	—	—	18 Cham	38 Cham	22 Mol*	—	—	RMF=Cham(Hanelt et al. 1992)
" ssp. chamaemoly	—	—	—	—	—	—	Mol	—	—	IK (Spain, Mediterranean) 1988
" " longicaulis Pastor et Valdés	—	—	—	—	—	—	Mol	—	—	IK (Spain, W.Mediterranean) 1988
" var. coloratum Batt.	—	—	—	—	—	—	21 Mol*	—	—	=tourneuxii
chamaespathum Boiss.	—	—	—	—	—	103 All	—	—	84	
chamarense M.Ivanova	—	—	—	—	—	—	Rhiz	Ret	—	IK (Siberia) 1965
chanetii Lévl.	89 Hap	—	—	—	—	—	—	—	—	=macrostemon
" Gandoger	—	—	—	—	—	—	—	—	—	NWF=?tenuissimum
changduense J.M.Xu	45 Rhiz	—	—	—	—	—	—	Ret	—	
charadzeae Tscholok.	—	—	—	—	—	—	Rhiz	Orei	—	IK (Caucasus) 1965 =gunibicum (Kud 1992)
charaulicum Fomin	—	—	109 Hap	—	68 Cod	—	—	—	—	USSR=?rupestre
chauvelii Boiss.	—	—	—	—	111 All	—	—	—	54a	=junceum ssp. junceum
chelotum Wend.	—	—	—	126 Meg	—	—	—	Meg	—	see also Fritsch 1996
chevsuricum Tscholok.	—	—	—	—	—	—	Rhiz	Orei	—	IK (Caucasus) 1965 =gunibicum (Kud 1992)
chienchuanense J.M.Xu	11 Brom	—	—	—	—	—	—	—	—	
chinense G.Don	81 Hap	—	36 Rhiz	—	—	—	—	Sacc	—	
" Maxim.	25 Rhiz	—	—	—	—	—	—	But	—	=tuberosum
chionanthum Boiss.	—	—	—	—	13 Mol	—	—	—	—	=zebdanense
chitralicum Wang et Tang	—	—	—	122 Meg	—	—	—	Meg	—	
chiwui Wang et Tang	50 Rhiz	—	—	—	—	—	—	Rhiz	—	
chloranthum Boiss.	—	—	—	—	39 Cod	—	14 Cod*	—	—	*=paniculatum
chloroneurum Boiss.	—	—	—	41 Scor	—	—	—	—	—	
chlorurum Boiss. et Hausskn.	—	—	—	—	50 Cod	—	—	—	—	=tauricolum
chodsha-bakirganicum [chodsa-] G.Gaffarov et I.Turakulov	—	190 Reg	—	—	—	—	Reg	Reg	—	Spelling mistake in IK. See Kh & Fr 1994
choriotepalum Wend.	—	—	—	47 Scor	—	—	—	—	—	
christii Janka	—	—	—	—	—	—	11 Rhiz	—	—	=lineare (Stearn 1978)
christophii Trautv.										see cristophii

chrysantherum Boiss. et Reut.	—	—	—	111 Mel	125 Mel	—	—	Mel	—	
chrysanthum Regel	74 Schoe	—	—	—	—	—	—	Orei	—	
chrysocephalum Regel	67 Rhiz	—	—	—	—	—	—	Orei	—	
chrysonemum Stearn	—	—	—	—	—	47 Scor	—	—	—	
ciliare Redouté	—	—	—	—	—	—	18a Mol*	—	—	*=subhirsutum ssp. subhirsutum
ciliatum Cirillo	—	—	—	—	8 Mol	—	18a Mol*	—	—	Turk=subhirsutum;*=s. " "
" C.Koch	—	—	—	—	112 All	—	—	—	57	Turk & BM=?jubatum
" Sims or Roth	—	—	—	—	8 Mol	—	18a Mol*	—	—	Turk=subhirsutum *= " ssp. subhirsutum
cilicicum auct. (non Boiss.)	—	—	—	—	—	90 All	—	—	—	=sphaerocephalon
" Boiss.	—	—	168 Porr	—	—	—	—	—	—	=rotundum
circassicum Kolak.	—	—	—	—	—	—	Mol	—	—	IK (Caucasus) 1955 =candolleanum (Kud 1992)
circinnatum Sieber	—	—	—	—	—	24 Mol	—	—	—	
circumflexum Wend.	—	—	—	52 Scor	—	—	—	Brevid	—	see Fritsch et al. 1998
cirrhosum Vandelli	—	—	—	—	—	71b Cod	—	—	—	=carinatum ssp. pulchellum
clarkei Hook.f.	25 Rhiz	—	—	—	—	—	24 Rhiz	But	—	Hook. rev. Stearn 1945; China & NWF=tuberosum
clathratum Ledeb.	—	—	8 Rhiz	—	—	—	—	Ret	—	
" auct. non Ledeb.	—	—	—	—	—	—	Rhiz	—	—	=ubsicolum (Friesen 1988)
clausum Vved.	—	58 Orei	+ Rhiz	—	—	—	—	Orei	—	
clusianum Retz.	—	—	—	—	—	30 Mol	18a Mol*	—	—	Eur=subhirsutum *= " ssp. subhirsutum
" auct.	—	—	—	—	—	—	18b Mol*	—	—	*= " ssp. subvillosum
coerulescens G.Don	90 Hap	—	129 Hap	—	—	—	54 Scor	—	—	=caeruleum (Stearn 1978)
coeruleum										see caeruleum
colchicifolium Boiss.	—	—	—	113 Mel	129 Mel	—	—	Mel	—	
collinum Guss.	—	—	—	—	—	—	?57c Cod	—	—	?=pallens ssp. siciliense (Stearn 1978)
collis-magni Kamelin	—	159 Acmo	—	—	—	—	—	Acmo	—	see Kh & Fr 1994
" Levichev non Kamelin	—	157 Acmo	—	—	—	—	—	Acmo	—	see Kh & Fr 1994; =tashkenticum
coloratum Spreng.	—	—	—	—	—	—	71b Cod	—	—	=carinatum ssp. pulchellum (Stearn 1978)
columnae Bubani	—	—	—	—	—	—	22 Mol*	—	—	=chamaemoly
commutatum Guss.	—	—	—	—	79 All	81 All	—	—	2	
compactum Thuill.	—	—	—	—	104 All	95 All	1 All*	—	85	=vineale (Eur=v. var. compactum)
complanatum (Fries) Boureau	—	—	—	—	—	—	63 Cod	—	—	=oleraceum (Stearn 1978)
condensatum Turcz.	64 Rhiz	—	67 Rhiz	—	—	—	—	Orei	—	

Column 1 Species of Allium & authorities (incl. synonyms)	2 Fl. China	3 Middle Asia	4 Fl. USSR (+= Fl. As. med.)	5 Fl. Iran	6 Fl. Turkey	7 Fl. Europ.	8 Others	9 Fritsch & Friesen	10 B. Math-ew	11 Accepted names for synonyms
confertum Jordan et Fourr.	—	—	—	—	—	23 Mol	25 Mol*	—	—	=roseum
confragosum Vved.	—	107 Scor	138 Hap	—	—	—	—	—	—	
confusum Halácsy	—	—	—	—	—	97b All	—	—	80b	=guttatum ssp. sardoum
congestum G.Don	—	—	—	—	—	—	—	Rhiz	—	=rubens
consanguineum Kunth	—	—	—	—	—	—	9 Rhiz	Orei	—	Hook. rev. Stearn 1945
controversum Schrader	—	—	—	—	—	75 All	—	—	—	=sativum var. ophioscorodon
" Costa	—	—	—	—	—	—	—	—	12	=pyrenaicum
convallarioides Grossh.	—	137 Cod	110 Hap	86 Cod	69 Cod	57 Cod	—	—	—	Turk & Eur perhaps=myrianthum
coppoleri Tineo	—	—	—	—	41 Cod	57a Cod	14 Cod* Pal 10 Cod	—	—	=pallens ssp. pallens (*=paniculatum)
cordifolium J.M.Xu	3 Ang	—	—	—	—	—	—	Ang	—	China=ovalifolium var. cordifolium
cornutum G.C.Clementi ex Vis.	—	—	—	—	—	20 Cep	—	Cep	—	NWF=triploid viviparous or bulbilliferous onion
costatovaginatum Kamelin et Levichev	—	166 Acmo	—	—	—	—	—	Acmo	—	IK (Tianshan) 1986; Kh & Fr 1994
cowanii Lindl.	—	—	—	—	—	—	23 Mol*	—	—	=?neapolitanum
crameri Asch. et Boiss.	—	—	—	—	—	—	31 Mel*	Mel	—	
cristatum Boiss.	—	—	—	—	112 All	100 All	—	—	57	=jubatum
cristophii Trautv. [christophii]	—	192 Kal	192 Mol + Mol	102 Acan	121 Acan	—	—	Kal	—	+=bodeanum
crystallinum Vved.	—	97 Crys	148 Porr	—	—	—	—	—	113	
cucullatum Wend.	—	—	—	135 Thaum	—	—	—	Thaum	—	
cupanii Rafin.	—	—	—	—	20 Brev	49 Scor	15 Cod*	—	—	
" ssp. anatolicum Stearn	—	—	—	—	20 Brev	—	49c Scor	—	—	Stearn 1978; Turk=cupani ssp. hirtovaginatum
" " cupanii	—	—	—	—	—	49a Scor	—	—	—	
" " hirtovaginatum (Kunth) Stearn	—	—	—	—	20 Brev	49b Scor	—	—	—	
cupuliferum Regel [-ll-]	—	183 Reg	223 Mol	—	—	—	—	Reg	—	see Kh & Fr 1994
" var. regelii (Trautv.) Kuntze	—	—	224 Mol	—	—	—	—	Reg	—	=regelii
curtum Boiss. et Gaill.	—	—	—	—	97 All	—	4c All* Pal 19 All	—	67	*=sphaerocephalon ssp. curtum

" ssp. aegyptiacum Täckholm et Drar	—	—	—	—	—	—	4c All*	—	67c	Egypt; *= " " "
" " curtum	—	—	—	—	—	—	4c All*	—	67a	Palestine; *= " " "
" " palaestinum Feinbr.	—	—	—	—	—	—	4c All*	—	67b	Palestine; *= " " "
" " typicum Feinbr.	—	—	—	—	—	—	4c All*	—	—	Lebanon; *= " " "
cyaneum Regel	36 Rhiz	—	—	—	—	—	—	Ret	—	
" var. brachystemon Regel	35 Rhiz	—	—	—	—	—	—	Ret	—	=sikkimense
cyathophorum Bur. et Franch.	13 Brom	—	—	—	—	—	Cyath	—	—	Hanelt & Fr 1994
" var. farreri Stearn	13a Brom	—	—	—	—	—	Cyath	—	—	Hanelt & Fr 1994
cyclospathum Freyn	—	—	—	—	109 All	—	—	—	58	=aucheri
cydni Schott et Kotschy ex Schott	—	—	—	—	11 Mol	—	—	—	—	=cassium
cyprium Brullo, Pavone et Salmeri	—	—	—	—	—	—	Cod	—	—	IK (Cyprus) 1993
cyrilli Ten.	—	—	—	—	124 Mel	107 Mel	29a Mel*	Mel	—	*=nigrum ssp. multibulbosum
czelghauricum Bordz.	—	—	—	—	4 Rhiz	—	—	Rhiz	—	
daghestanicum Grossh.	—	—	78 Rhiz	—	—	—	—	Orei	—	
dalmaticum A.Kerner ex Janchen	—	—	—	—	106 All	97c All	—	—	80c	=guttatum ssp. dalmaticum
damascenum Feinbr.	—	—	—	—	—	—	—	—	61	Syria
daninianum Brullo, Pavone et Salmeri	—	—	—	—	—	—	Cod	—	—	IK (Middle East) 1996
darwasicum Regel [darvasicum]	—	189 Reg	221 Mol	—	—	—	—	Reg	—	see Kh & Fr 1994
dasyphyllum Vved.	—	153 Acmo	199 Mol	—	—	—	—	Acmo	—	
dauricum Friesen	—	—	—	—	—	—	Rhiz	Rhiz	—	IK (Siberia; China) 1987
davisianum Feinbr.	—	—	—	—	101 All	—	—	—	—	=phanerantherum ssp. phanerantherum
" " emend. Kollm.	—	—	—	—	—	—	Pal 18 All	—	60	= " " "
decaisnei C.Presl	—	—	—	—	—	—	Pal 13a Cod	—	—	
deciduum Özhatay et Kollm.	—	—	—	—	59 Cod	—	—	—	—	
" ssp. deciduum	—	—	—	—	59 Cod	—	—	—	—	
" " retrorsum Özhatay et Kollm.	—	—	—	—	59 Cod	—	—	—	—	
decipiens Fisch.	—	—	+ Mol	—	—	—	—	—	—	
" var. integrifolium Parsa	—	80 Cep	—	19 Cep	—	—	—	—	—	=oschaninii
" Fisch. ex Schult. et Schult.f.	—	—	—	—	130 Mel	108 Mel	—	Mel	—	
" " ex Roem. et Schult.	96 Mol	—	202 Mol	—	—	—	—	—	—	
" " " " var. latissimum Lipsky	—	—	205 Mol	—	—	—	—	—	—	= grande
" Vved. p.p.	—	141 Mel	—	—	—	—	—	—	—	=tulipifolium
delicatulum Siev. ex Schult. et Schult.f. [ex Roem. et Schult.] [delicatum]	—	123 Caer	125 Hap	—	—	45 Scor	—	—	—	

Column 1 Species of Allium & authorities (incl. synonyms)	2 Fl. China	3 Middle Asia	4 Fl. USSR (+= Fl. As. med.)	5 Fl. Iran	6 Fl. Turkey	7 Fl. Europ.	8 Others	9 Fritsch & Friesen	10 B. Mathew	11 Accepted names for synonyms
deltoide-fistulosum S.O.Yu, S.Lee et W.T.Lee	—	—	—	—	—	—	—	Sacc	—	IK (Korea) 1981
densiflorum Hampe	—	—	—	—	—	—	—	—	80b	=guttatum ssp. sardoum
dentiferum Webb et Berthelot	—	—	—	—	—	—	Cod BPS 14 Cod*	—	—	BPS 1991 *=paniculatum
dentigerum Prokh.	32 Rhiz	—	—	—	—	—	—	Caes	—	
derderianum sensu Grossh.	—	—	—	—	127 Mel	—	—	—	—	=woronowii
" Regel	—	—	188 Mol	95 Acan	—	—	—	Acan	—	
descendens auct. (non L.) p.p.	—	—	—	—	105 All	96 All	—	—	81	=amethystinum
" " " "	—	—	—	—	101 All	—	Pal 18 All	—	60	=phanerantherum (Turk= p. ssp. phanerantherum)
" "	—	—	159 Porr	—	—	—	—	—	—	=sphaerocephalon
" L.	—	—	—	—	95 All	—	—	—	62a	= " ssp. sphaerocephalon
" var. tenuifolium Miscz. ex Grossh.	—	—	160 Porr	—	—	—	—	—	—	=regelianum
deserticolum M.Pop.	16 Rhiz	65 Orei	+ Rhiz	—	—	—	—	Camp	—	=teretifolium (exc. China)
deserti-syriaci Feinbr.	—	—	—	—	—	—	—	—	96	Iraq
desertorum Forssk.	—	—	—	—	—	—	14 Cod* Pal 11 Cod	—	—	*= paniculatum
diaphanum Janka	—	—	32 Rhiz	—	—	—	—	Camp	—	=inderiense
dictyoprasum C.A.Mey. ex Kunth	—	—	152 Porr	—	115 All	—	Pal 23 All	—	101	
" var. virescens Grossh.	—	—	—	—	—	—	—	—	101	=dictyoprasum
dictyoscordum Vved.	—	86 All	150 Porr	63 All	—	—	—	—	50	
dilatatum Zahar.	—	—	—	—	—	98 All	—	—	80d	BM=guttatum ssp. dilatatum Eur perhaps=ssp. of guttatum
dilutum Stapf	—	—	—	106 Mel	135 Mel	—	—	Mel	—	=noëanum; RMF=perhaps good sp.
dinsmorei Rech.f.	—	—	—	—	—	—	Pal 2 Mol	—	—	=trifoliatum var. sterile
dioscoridis Sibth. et Sm.	—	—	228 Nect	—	—	—	—	—	—	
" auct. (non Sibth. et Sm.)	—	—	—	—	—	Nect	—	—	—	=Nectaroscordum siculum ssp. bulgaricum
djimilense Boiss. ex Regel	—	—	—	—	44 Cod	—	—	—	—	
dodecadontum Vved.	—	164 Acmo	+ Mol	—	—	—	—	Acmo	—	see Kh & Fr 1994
dodecanesi Karavokyrou et Tzan.	—	—	—	—	—	—	Cod	—	—	IK (Greece) 1994

dolichomischum Vved.	—	29 Camp	30 Rhiz	—	—	—	—	Camp	—	
dolichostylum Vved.	—	31 Camp	31 Rhiz	9 Rhiz	—	—	—	Camp	—	
dregeanum Kunth	—	—	—	—	—	—	8 All*	—	47	
drepanophyllum Vved.	—	21 Camp	25 Rhiz	—	—	—	—	Camp	—	
drobovii Vved.	—	17 Ret	5 Rhiz	—	—	—	—	Ret	—	
drusorum Feinbr.	—	—	—	—	—	—	—	—	37	Syria
dshambulicum Pavl.	—	132 Minu	+ Hap	—	—	—	—	—	—	=parvulum
dshungaricum Vved.	—	60 Orei	+ Rhiz	—	—	—	—	Orei	—	nom. inval. (NWF)
dumetorum Feinbr. et Szelubsky	—	—	—	—	—	—	Pal 24 Mel	Mel	—	Pal=nigrum
duriaeanum J. Gay [durieanum; sphalm. durtaeanum]	—	—	—	—	—	—	9 All*	—	48	=baeticum
dyctioprasum (sphalm.)	—	—	—	—	115 All	—	Pal 23 All	—	—	=dictyoprasum
ebusitanum Font Quer	—	—	—	—	—	95 All	—	—	65	Eur perhaps=vineale
ecornutum F.O.Khassanov et I.I.Maltsev	—	146 Meg	—	—	—	—	—	Meg	—	IK (Uzbekistan) 1988; Kh & Fr 1994 =jesdianum ssp. angustitepalum
edentatum Y.P.Hsu	—	—	—	—	—	—	?Rhiz	Caes	—	IK (China) 1987
eduardii Stearn	17 Rhiz	4 Ret	+ Rhiz	—	—	—	—	Ret	—	
effusum Boiss.	—	—	—	—	64 Cod	—	—	—	—	=stamineum
eginense Freyn p.p.	—	—	—	—	125 Mel	—	—	Mel	—	Turk=chrysantherum
" " "	—	—	—	—	130 Mel	—	—	—	—	=decipiens
eivissanum Garbari et Miceli	—	—	—	—	—	—	Brev	—	—	IK (Balearic Is.) 1987
elatum Regel	—	176 Comp	216 Mol	128 Meg	—	—	—	Comp	—	=macleanii (exc. USSR)
elburzense Wend.	—	—	—	105 Acan	—	—	—	Acan	—	
eldivanense Özhatay	—	—	—	—	98a All	—	—	—	72	Turk vol. 10
elegans Drob.	—	120 Caer	131 Hap	—	—	—	—	—	—	
elegantulum Kitag.	46 Rhiz	—	—	—	—	—	—	Ten	—	=tenuissimum
ellisii Hook.f.	—	—	—	104 Acan (cult.)	—	—	—	Kal	—	
emarginatum Rech.f.	—	—	—	—	115 All	—	Pal 23 All	—	101	=dictyoprasum (Turk=?dict.)
enginii Özhatay et B.Mathew	—	—	—	—	—	—	—	—	46	S.Turkey
erdelii Zucc.	—	—	—	—	—	—	17 Mol* Pal 4 Mol	—	—	
" var. hirtellum Opphr.	—	—	—	—	—	—	17 Mol* Pal 4 Mol	—	—	=erdelii
" " lasiophyllum Nábelek	—	—	—	—	—	—	Pal 5 Mol	—	—	=qasyunense
" " micranthum Opphr.	—	—	—	—	—	—	Pal 5 Mol	—	—	= "
" " roseum Boiss.	—	—	—	—	—	—	17 Mol* Pal 4 Mol	—	—	=erdelii

Column 1 Species of Allium & authorities (incl. synonyms)	2 Fl. China	3 Middle Asia	4 Fl. USSR (+= Fl. As. med.)	5 Fl. Iran	6 Fl. Turkey	7 Fl. Europ.	8 Others	9 Fritsch & Friesen	10 B. Math-ew	11 Accepted names for synonyms
erectum G.Don	—	—	—	—	—	—	?	—	—	See Stearn 1978 p. 175
eremoprasum Vved.	—	139 Vved	136 Hap	—	—	—	—	—	—	
ericetorum Thore	—	—	—	—	—	8 Rhiz	—	Orei	—	
eriocoleum Vved.	—	14 Ret	+ Rhiz	—	—	—	—	Ret	—	
eriophyllum Boiss.	—	—	—	90 Mol	14 Mol	—	—	—	—	Turk=longisepalum
" var. eriophyllum	—	—	—	90 Mol	—	—	—	—	—	
" " laceratum (Boiss. et Noë) Wend.	—	—	—	90 Mol	—	—	—	—	—	
erubescens C.Koch	—	—	167 Porr	73 All	89a All	—	—	—	44	Turk vol. 10
erythraeum Griseb.	—	—	—	—	s.n. Scor	51 Scor	—	Scor	—	BPST 1994; Eur=obtusiflorum
										Turk may=rubellum
esfandiarii Matin	—	—	—	—	—	—	—	—	27	Iran
euboicum Rech. f.	—	—	—	—	—	56c Cod	—	—	—	=paniculatum ssp. euboicum
eugenii Vved.	—	201 Brevic	184 Mol	—	—	—	—	Brevic	—	
eusperma Airy-Shaw	86 Hap	—	—	—	—	—	—	—	—	
exiguiflorum Hayek et Siehe	—	—	—	87 Cod	69 Cod	—	—	—	—	=myrianthum
exile Boiss. et Orph.	—	—	—	—	48 Cod	—	—	—	—	=sipyleum
eximium Siehe	—	—	—	—	134 Mel	—	—	—	—	=aschersonianum
exsertum Baker	—	—	—	—	—	—	—	Sacc	—	=chinense
fallax Schult. et Schult.f.	—	—	—	—	—	3 Rhiz	—	Rhiz	—	Eur=senescens ssp. montanum
										NWF=lusitanicum
" Roem. et Schult.	51 Rhiz	—	44 Rhiz	—	—	—	—	—	—	=senescens
faniae Stearn	—	—	—	—	—	—	—	—	58	=aucheri; see also Stearn 1978
farctum Wend.	—	—	—	18 Cep	—	—	—	Cep	—	
farreri Stearn	13a Brom	—	—	—	—	—	—	—	—	=cyathophorum var. farreri
farsistanicum Gand.	—	—	—	90 Mol	—	—	—	—	—	=eriophyllum var. eriophyllum
fasciculatum Rendle	9 Brom	—	—	—	—	—	—	—	—	
fastigiatum Cand.	—	—	—	—	64 Cod	—	—	—	—	=?stamineum
favosum Zahar.	—	—	—	—	—	60 Cod	—	—	—	
fedschenkoanum Regel [fedt-]	—	72 Ann	81 Rhiz	16 Schoe	—	—	—	Ann	—	USSR=monadelphum
" auct.	—	—	+ Rhiz	—	—	—	—	—	—	=kaufmannii
fedtschenkoi Nábelek	—	—	—	109 Mel	—	—	—	Mel	—	
feinbergii Oppenh.	—	—	—	—	—	—	Scor	—	—	IK (Palestine) 1940 (not in Fl. Pal.)

ferganicum Vved.	—	99 Mult	143 Porr	—	—	—	—	—	79	
ferrinii Pamp.	—	—	—	—	111 All	—	—	—	54a	=junceum ssp. junceum
fethiyense Özhatay et B.Mathew	—	—	—	—	—	—	—	—	103	SW.Turkey
fetisowii Regel [fetissovii]	97 Mol	162 Acmo	198 Mol	—	—	—	—	Acmo	—	see Kh & Fr 1994
fibriferum Wend.	—	—	—	123 Meg	—	—	—	Acmo	—	see Kh & Fr 1994
fibrosum Regel	—	114 Avu	112 Hap	38 Scor	—	—	—	—	—	
filidens Regel	—	88 Cost	149 Porr	62 All	—	—	—	—	107	
filidentiforme Vved. ex Kaschtsch.	—	90 Cost	+ Porr	—	—	—	—	—	108	
[et Nikit.]										
filifolium Freyn et Sint.	—	—	—	—	56 Cod	—	—	—	—	=pseudoflavum
" Regel	—	—	64 Rhiz	13 Rhiz	—	—	—	—	—	
" "	—	53 Orei	+ Rhiz	—	—	—	—	Orei	—	=kokanicum
fimbriatum Schischkin	—	—	98 Hap	—	21 Brev	—	—	—	—	USSR=lacerum; Turk=callidictyon
firmotunicatum Fom.	—	—	157 Porr	—	—	79 All	—	—	13	BM (& Eur?)=atroviolaceum
" var. album Grossh.	—	—	174 Porr	—	—	—	—	—	6	=leucanthum
" f. album (Grossh.) Grossh.	—	—	—	—	—	—	—	—	6	= "
fischeri Regel	17 Rhiz	—	7 Rhiz	—	—	—	—	Ret	—	=eduardii (exc. USSR)
			+ Rhiz							
fistulosum L.	76 Schoe	—	88 Phyll	—	—	21 Cep	11 Schoe*	Cep	—	
			(cult.)							
" Ledeb.	—	—	87 Phyll	—	—	—	—	—	—	=altaicum
flavellum Vved.	—	30 Camp	+ Rhiz	—	—	—	—	Camp	—	
flavescens Bess.	—	—	39 Rhiz	—	—	6a Rhiz	—	Rhiz	—	Eur=albidum ssp. albidum
" var. ammophilum (Heuffel)	—	—	—	—	—	—	6a Rhiz	—	—	=albidum ssp. alb. (Stearn 1978)
Zahar.										
flavidum Ledeb.	21 Rhiz	5 Ret	10 Rhiz	—	—	—	—	Ret	—	
" auct. non Ledeb.	—	—	—	—	—	—	—	—	—	=alaschanicum, q.v.
flavovirens Regel	20 Rhiz	—	9 Rhiz	—	—	—	—	Ret	—	=leucocephalum
flavum auct. (non L.) p.p.	—	—	101 Hap	—	56 Cod	—	14 Cod*	—	—	*=paniculatum USSR & Turk=pseudoflavum
" " " p.p.	—	—	102 Hap	—	—	—	—	—	—	=carinatum ssp. pulchellum (Stearn 1978)
" Salisb.	—	—	—	—	—	—	33 Mol	—	—	=moly (Stearn 1978)
" L.	—	—	—	—	55 Cod	68 Cod	14 Cod*	—	—	*=paniculatum
" ssp. flavum	—	—	—	—	55 Cod	68a Cod	—	—	—	
" " pulchellum K.Richter	—	—	—	—	—	—	71b Cod	—	—	=carinatum ssp. pulchellum (Stearn 1978)

Column 1 Species of Allium & authorities (incl. synonyms)	2 Fl. China	3 Middle Asia	4 Fl. USSR (+= Fl. As. med.)	5 Fl. Iran	6 Fl. Turkey	7 Fl. Europ.	8 Others	9 Fritsch & Friesen	10 B. Math-ew	11 Accepted names for synonyms
flavum ssp. tauricum (Besser ex Reichb.) K.Richter [or Stearn]	—	—	—	—	55 Cod	68b Cod	—	—	—	Turk=flavum ssp. tauricum var. tauricum
" var. humile Regel	—	—	—	—	66 Cod	—	—	—	—	=pictistamineum p.p.
" " minus Boiss. p.p.	—	—	—	—	66 Cod	—	—	—	—	"
" " minus Boiss. p.p.	—	—	—	—	55 Cod	—	—	—	—	
" " pilosum Kollm. et Koyuncu	—	—	—	—	55 Cod	—	—	—	—	=flavum ssp. tauricum var. pilosum
" " pulchellum (G.Don) Regel or Ledeb.	—	—	102 Hap	—	58 Cod	71b Cod	—	—	—	Turk & Eur=carinatum ssp. pulchellum; USSR=pulchellum
" " purpurascens Mertens et Koch	—	—	—	—	—	—	71b Cod	—	—	=carinatum ssp. pulchellum (Stearn 1978)
" " ruthenicum Lang	—	—	—	—	55 Cod	—	—	—	—	=flavum ssp. tauricum var. tauricum
" " sordideroseum Czern.	—	—	—	—	—	—	68b Cod	—	—	=flavum ssp. tauricum (Stearn 1978)
" " tauricum Besser ex Reichb.	—	—	102 Hap	—	55 Cod	68b Cod	—	—	—	Turk & Eur= " " " var. tauricum; USSR=pulchellum
flexum Waldst. et Kit.	—	—	—	—	58 Cod	—	—	—	—	=carinatum ssp. carinatum
" var. capsuliferum Koch	—	—	—	—	—	—	71b Cod	—	—	= " " pulchellum (Stearn 1978)
flexuosum d'Urv.	—	—	—	—	53 Cod	—	—		—	=staticiforme
foliosum Clarion ex DC.	—	—	—	—	—	18 Schoe	—	Schoe	—	=schoenoprasum (Eur=var. alpinum or var. or ssp. sibiricum)
fominianum Miscz.	—	—	164 Porr	—	92 All	—	—	—	25	Turk & BM=gramineum
fominii Miscz. ex Grossh.	—	—	164 Porr	—	92 All	—	—	—	25	USSR=fominianum; Turk & BM =gramineum
fontanesii J.Gay	—	—	—	—	—	—	14 Cod*	—	—	=paniculatum
forrestii Diels	43 Rhiz	—	—	—	—	—	—	Ret	—	
fragrans Vent.	—	—	—	—	—	Noth	—	—	—	Eur=Nothoscordum inodorum; =N. gracile (Stearn 1986) or N. borbonicum (Ravenna 1991)
" Cirillo [ex Ten.]	—	—	—	—	—	—	29a Mel*	Mel	—	RMF=nigrum, *=nigrum ssp. multibulbosum
franciniae Brullo et Pavone	—	—	—	—	—	—	Scor	—	—	IK (Sicily) 1983
freynianum Bornm.	—	—	—	—	56 Cod	—	—	—	—	=pseudoflavum
freynii Bornm.	—	—	—	—	56 Cod	—	—	—	—	"
frigidum Boiss. et Heldr.	—	—	—	—	31 Scor	46 Scor	—	—	—	

fritschii Khassanov et Iengalycheva	—	94 Cost	—	—	—	—	—	—	—	
frivaldszkyanum Kunze [-dz-]	—	—	—	—	—	—	—	—	80b	=guttatum ssp. sardoum Spelling mistake in IK
funckiaefolium Hand.-Mazz.	4 Ang	—	—	—	—	—	—	Ang	—	
fuscoviolaceum Fom.	—	—	158 Porr	—	100 All	—	—	—	66	
fuscum Waldst. et Kit.	—	—	—	—	38 Cod	56b Cod	14 Cod* Pal 9 Cod	—	—	=paniculatum ssp. fuscum (*=paniculatum)
" auct.	—	—	105 Hap	—	—	—	—	—	—	=paniculatum
fussii Kerner	—	—	—	—	—	—	Cod	—	—	=paniculatum ssp. fuscum (Fl.Român.)
gaditanum Pérez Lara ex Willk.	—	—	—	—	—	97b All	2 All*	—	80b	Eur & BM=guttatum ssp. sardoum; *=guttatum
" var. bulbiferum Coutinho	—	—	—	—	—	—	—	—	85	=vineale
gageanum W.W.Smith	9 Brom	—	—	—	—	—	29 Mol	—	—	=fasciculatum; Hook. rev. Stearn 1945
galanthum Kar. et Kir.	79 Cep	77 Cep	89 Cep	—	—	—	—	Cep	—	
gasparrinii Guss.	—	—	—	—	—	—	—	—	1	=ampeloprasum
gayi Boiss.	—	—	—	—	10 Mol	—	30 Mel*	—	—	*=orientale
getulum Batt. et Trab.	—	—	—	—	—	—	7 All*	—	1	=ampeloprasum
giganteum Regel	—	175 Comp	217 Mol	127 Meg	—	—	—	Comp	—	see Kh & Fr 1994
" Wend.	—	—	—	128 Meg	—	—	—	—	—	=macleanii
gilgiticum Wang et Tang	—	—	—	—	—	—	30 Mol	—	—	Hook. rev. Stearn 1945
gillii Wend.	—	—	—	50 Scor	—	—	—	—	—	
glaciale Vved.	—	48 Orei	58 Rhiz	—	—	—	—	Orei	—	
glaucum Schrad.	—	—	44 Rhiz	—	—	—	—	Rhiz	—	=senescens
globosum Boiss. p.p.	—	—	71 Rhiz	—	—	—	—	—	—	=saxatile
" M.Bieb. [ex Redouté]	63 Rhiz	61 Orei	72 Rhiz	—	—	4 Rhiz	—	Orei	—	Eur=saxatile
" var. albidum Regel	—	—	66 Rhiz	—	—	—	—	—	—	=tianschanicum
" " ochroleucum Boiss.	—	—	70 Rhiz	—	—	—	—	—	—	=marschallianum
" " " Regel	—	57 Orei	—	—	—	—	—	—	—	=petraeum
" " saxatile Schmalh. p.p.	—	—	70 Rhiz	—	—	—	—	—	—	=marschallianum
" " " " "	—	—	71 Rhiz	—	—	—	—	—	—	=saxatile
" forma dilute-roseum Kryl.	—	60 Orei	71 Rhiz	—	—	—	—	—	—	Asia=dshungaricum, USSR=saxatile
" auct.	—	—	+ Rhiz	—	—	—	—	—	—	=dshungaricum & talassicum
glomeratum Prokh.	87 Hap	124 Caer	126 Hap	—	—	—	—	—	—	
glumaceum Boiss. et Hausskn.	—	—	—	—	46 Cod	—	—	—	—	
gmelinianum Miscz. ex Grossh.	—	—	72 Rhiz	—	—	—	—	Orei	—	=globosum

Column 1 Species of Allium & authorities (incl. synonyms)	2 Fl. China	3 Middle Asia	4 Fl. USSR (+= Fl. As. med.)	5 Fl. Iran	6 Fl. Turkey	7 Fl. Europ.	8 Others	9 Fritsch & Friesen	10 B. Mathew	11 Accepted names for synonyms
goloskokovii Vved.	—	56 Orei	+ Rhiz	—	—	—	—	Orei	—	
gomphrenoides Boiss. et Heldr.	—	—	—	—	—	99 All	—	—	52	
gorumsense Boiss.	—	—	—	—	108 All	—	—	—	83	
goulimyi Tzan.	—	—	—	—	—	—	Scor	—	—	IK (Greece) 1983
govanianum Wall. ex Baker	—	—	—	15 Rhiz	—	—	—	—	—	=humile
gracile Albov	—	—	76 Rhiz	—	—	—	—	Orei	—	=albovianum
" Aiton	—	—	—	—	—	—	—	—	—	=Nothoscordum gracile (Stearn 1986)
gracilescens Somm. et Lev.	—	—	166 Porr	—	—	—	—	—	38	BM=?ponticum
gracillimum Vved.	—	11 Ret	+ Rhiz	—	—	—	—	Ret	—	
graecum d'Urv.	—	—	—	—	9 Mol	—	18a Mol*	—	—	*=subhirsutum ssp. subhirsutum Turk=trifoliatum
" sensu J.Gay	—	—	—	—	10 Mol	—	—	—	—	=gayi
" auct.	—	—	—	—	—	—	30 Mel*	—	—	=orientale
gramineum C.Koch	—	—	—	—	92 All	—	—	—	25	
" var. ampeloprasoides (Miscz. ex Grossh.) Grossh.	—	—	—	—	92 All	—	—	—	25	=gramineum
grande Lipsky	—	—	205 Mol	s.n. Meg	—	—	—	Meg	—	
grandiflorum Lam.	—	—	—	—	—	—	15 Rhiz; Nark	—	—	=narcissiflorum (Stearn 1978); RMF=Nark
grayi Regel	89 Hap	—	—	—	—	—	Ohwi	—	—	Chin=macrostemon
gredense Rivas Mateos [Goday]	—	—	—	—	—	18 Schoe	—	Schoe	—	=schoenoprasum
greuteri Brullo et Pavone	—	—	—	—	—	—	Brev	—	—	IK (Libya) 1983
griffithianum Boiss.	—	116 Avu	121 Hap	48 Scor	—	—	—	—	—	
grimmii Regel	14 Rhiz	65 Orei	74 Rhiz	—	—	—	—	Camp?	—	=teretifolium
grisellum J.M.Xu	80 Hap	—	—	—	—	—	—	—	—	
grosii Font Quer	—	—	—	—	—	48 Scor	—	—	—	
guanxianense J.M.Xu	—	—	—	—	—	—	?Brom	—	—	IK (China) 1993
gubanovii Kamelin	—	—	—	—	—	—	Rhiz	Ret	—	IK (USSR) 1980
guicciardii Heldr. [-ccar-]	—	—	—	—	—	68a Cod	—	—	—	=flavum ssp. flavum
gulczense [gultschense] O.Fedtsch. [B.Fedtsch.]	—	152 Acmo	208 Mol	—	—	—	—	Acmo	—	RMF=backhousianum See OF 1906; Kh & Fr 1994
gunibicum Miscz. ex Grossh.	—	—	77 Rhiz	—	—	—	—	Orei	—	

gusaricum Regel [guzaricum]	—	19 Camp	23 Rhiz	—	—	—	—	Camp	—	
guttatum Steven	—	—	155 Porr	—	106 All	97 All	2 All*	—	80	see also Stearn 1978
" ssp. dalmaticum (A.Kerner ex Janchen) Stearn	—	—	—	—	106 All	97c All	—	—	80c	
" " dilatatum (Zahar.) B.Mathew	—	—	—	—	—	—	—	—	80d	Crete
" " guttatum	—	—	—	—	106 All	97a All	—	—	80a	
" " sardoum (Moris) Stearn	—	—	—	—	106 All	97b All	—	—	80b	
" " tenorei (Parl.) Soldano	—	—	—	—	—	—	All	—	—	Soldano 1994
guzaricum Regel										see gusaricum
gypsaceum M.Pop. et Vved.	—	180 Pop	181 Mol	—	—	—	—	Pop	—	see Kh & Fr 1994
gypsodictyum Vved.	—	93 Cost	+ Porr	—	—	—	—	—	51	
haemanthoides Boiss. et Reut. ex Regel	—	—	—	97 Acan	—	—	—	Acan	—	
" var. lanceolatum Boiss.	—	—	188 Mol	—	—	—	—	—	—	=derderianum
halleri G.Don	—	—	—	—	76 All	—	Pal 15 All	—	1	=ampeloprasum
hamrinense Hand.-Mazz.	—	—	—	67 All	—	—	—	—	95	
haneltii Khassanov et R.M.Fritsch	—	—	—	—	—	—	Brevid	—	—	(Uzbekistan); Fritsch et al. 1998
haussknechtii Nábelek	—	—	—	113 Mel	129 Mel	—	—	Mel	—	=colchicifolium
hedgei Wend.	—	—	—	60 All	—	—	—	Brevid	99	see Fritsch et al. 1998
heldreichii Boiss.	—	—	—	—	—	104 All	—	—	55	
helicophyllum Vved.	—	197 Kal	182 Mol	132 Kal	—	—	—	Kal	—	
hemisphaericum (Sommier) S.Brullo	—	—	—	—	—	—	All	—	—	=ampeloprasum var. hemisphaericum Sommier (IK Supp. 20)
henryi C.H.Wright	38 Rhiz	—	—	—	—	—	—	Ret	—	
herderianum Regel	68 Rhiz	—	—	—	—	—	—	Orei	—	
heteronema Wang et Tang	39 Rhiz	—	—	—	—	—	—	Ret	—	
hexaceras Vved.	—	204 Acaul	+ Mol	—	—	—	—	Acaul	—	see Kh & Fr 1994
hierochuntinum Boiss.	—	—	—	—	—	—	5 All* Pal 20 All	—	88	*=ascalonicum
hierosolymorum Regel	—	—	—	—	—	—	Pal 2 Mol	—	—	perhaps=trifoliatum var. hirsutum
himalayense Regel	—	—	—	57 Scor	—	—	—	—	—	=stocksianum
hindukuschense Kamelin et Seisums	—	—	—	—	—	—	—	Kal	—	IK (Afghanistan) 1996 RMF=good sp.
hirsutum Lam.	—	—	—	—	8 Mol	—	18a Mol*	—	—	=subhirsutum [ssp. subhirsutum]
" Zucc.	—	—	—	—	—	—	18a Mol* Pal 2 Mol	—	—	*= " " " Pal=trifoliatum var. hirsutum

Column 1 Species of Allium & authorities (incl. synonyms)	2 Fl. China	3 Middle Asia	4 Fl. USSR (+= Fl. As. med.)	5 Fl. Iran	6 Fl. Turkey	7 Fl. Europ.	8 Others	9 Fritsch & Friesen	10 B. Math-ew	11 Accepted names for synonyms
hirtifolium Boiss.	—	148 Meg	—	120 Meg	140 Mel	—	—	Meg	—	Asia & RMF=stipitatum; see also Fritsch 1996
hirtovaginatum Kunth	—	—	—	—	20 Brev	49b Scor	15 Cod*	—	—	Eur & Turk=cupanii ssp. hirtovaginatum; *=cupanii
hirtovaginum Cand.	—	—	—	—	65 Cod	72 Cod	—	—	—	
hissaricum Vved.	—	188 Reg	+ Mol	—	—	—	—	Reg	—	see Kh & Fr 1994
hoeltzeri Regel	61 Rhiz	—	65 Rhiz	—	—	—	—	—	—	=caricoides
" "	—	53 Orei	+ Rhiz	—	—	—	—	Orei	—	=kokanicum
hollandicum R.M.Fritsch	—	—	—	—	—	—	Meg	Meg	—	cult.; Fritsch 1993
holmense Miller	—	—	—	—	—	—	7 All*	—	—	=ampeloprasum
hookeri Thwaites	7 Brom	—	—	—	—	—	—	—	—	
" var. muliense Airy-Shaw	7a Brom	—	—	—	—	—	—	—	—	
hopeiense Nak.	62 Rhiz	—	—	—	—	—	—	—	—	=longistylum
horvatii Lovrić	—	—	—	—	—	5 Rhiz	—	Orei	—	
huber-morathii Kollm., Özhatay et Koyuncu	—	—	—	—	63 Cod	—	—	—	—	
hugonianum Rendle	36 Rhiz	—	—	—	—	—	—	Ret	—	=cyaneum
× humbertii Maire	—	—	—	—	—	—	18b Mol*	—	—	=subhirsutum ssp. subvillosum
humile Kunth	—	—	—	15 Rhiz	—	—	—	—	—	
" var. trifurcatum Wang et Tang	8 Brom	—	—	—	—	—	—	—	—	=trifurcatum
hymenorrhizum Ledeb.	—	46 Orei	56 Rhiz	12 Rhiz	—	13 Rhiz	—	Orei	—	
" var. tianschanicum Regel	—	—	66 Rhiz	—	—	—	—	—	—	=tianschanicum
hymettium Boiss. et Heldr.	—	—	—	—	—	73 Cod	—	—	—	
hypsistum Stearn	—	—	—	—	—	—	?Rhiz	Ret	—	IK (Nepal) 1960
iatasen Lévl.	89 Hap	—	—	—	—	—	—	—	—	=macrostemon
idzuense Hara	—	—	—	—	—	—	Rhiz	Schoe	—	IK (Japan) 1974; Friesen 1996=maximowiczii
ilgazense Özhatay	—	—	—	—	112a All	—	—	—	56	Turk vol. 10
iliense Regel	—	186 Reg	225 Mol	137 Reg	—	—	—	Reg	—	
illyricum Jacquin	—	—	—	—	—	—	25 Mol*	—	—	=roseum
inaequale Janka	—	102 Scor	95 Hap	—	—	41 Scor	—	—	—	
incarnatum Hornem.	—	—	—	—	—	—	23 Mol	—	—	=roseum (Stearn 1978)

incensiodorum Radić	—	—	—	—	—	—	Rhiz	—	—	Radić 1989
incisum Fom.	—	—	98 Hap	—	22 Brev	—	—	—	—	USSR=lacerum
inconspicuum Vved.	—	20 Camp	24 Rhiz	—	—	—	—	Camp	—	
incrustatum Vved.	—	111 Scor	+ Hap	—	—	—	—	—	—	
inderiense Fisch. ex Bge. [ex Schult. et Schult. f.]	—	32 Camp	32 Rhiz	—	—	17 Rhiz	—	Camp	—	
inodorum Aiton	—	—	—	—	—	—	Mol	—	—	=neapolitanum; see Stearn 1986
inops Vved.	—	127 Caer	123 Hap	—	—	—	—	—	—	
insubricum Boiss. et Reut.	—	—	—	—	—	16 Rhiz	Nark	—	—	RMF=Nark
insufficiens Vved.	—	171 Acmo	200 Mol	—	—	—	—	Acmo	—	see Kh & Fr 1994
integerrimum Zahar.	—	—	—	—	—	102 All	—	—	63	
intermedium DC.	—	—	—	—	38 Cod	56a Cod	14 Cod* Pal 9 Cod	—	—	=paniculatum ssp. paniculatum; (*=paniculatum)
" ssp. podolicum (Blocki ex Racib. et Szafer) Zapal.	—	—	—	—	—	62 Cod	—	—	—	=podolicum
inutile Makino	—	—	—	—	—	—	Ohwi	—	—	Ohwi=Nothoscordum inutile (Makino) Kitam.; NWF=A.neriniflorum?
involucratum Welw. ex Coutinho	—	—	—	—	—	—	2 All*	—	80b	*=guttatum; BM=g. ssp. sardoum
ionandrum Wend.	—	—	—	54 Scor	—	—	—	Brevid	—	see Fritsch et al. 1998
ionicum Brullo et Tzan.	—	—	—	—	—	—	Cod or Scor	—	—	IK (Greece) 1994
iranicum (Wend.) Wend.	—	—	—	—	—	—	—	—	5	=ampeloprasum ssp. iranicum Iran, Iraq
isauricum Hub.-Morr. et Wend.	—	—	—	—	16 Mol	—	—	—	—	
isfairamicum [isphair-] O.Fedtsch. [B.Fedtsch.]	—	176 Comp	216 Mol	—	—	—	—	Comp	—	USSR=elatum, Asia & RMF=?macleanii; see OF 1906
jacquemontii Regel	18 Rhiz	—	—	—	—	—	16 Rhiz	—	—	China=przewalskianum; Hook. rev. Stearn 1945=?stoliczkii
" Kunth	—	—	117 Hap	45 Scor	—	—	—	—	—	
" Vved.	—	115 Avu	—	46 Scor	—	—	—	—	—	=pamiricum
jajlae Vved.	—	—	169 All	—	89 All	87c All	—	—	41b	BM=rotundum ssp. jajlae; Turk & Eur=scorodoprasum ssp. jajlae
jaluanum Nakai	—	—	—	—	—	—	Rhiz	Orei	—	=condensatum (Friesen 1988)
janthinum Freyn	—	—	—	—	33 Scor	—	—	—	—	=tchihatschewii
japonicum Regel	84 Hap	—	—	—	—	—	—	—	—	=thunbergii
jaxarticum Vved.	—	106 Scor	+ Hap	—	—	—	—	—	—	
jeholense Franch.	62 Rhiz	—	—	—	—	—	—	—	—	=longistylum
jenischianum Regel p.p.	—	—	—	106 Mel	135 Mel	—	—	Mel	—	Iran & Turk=noëanum

Column 1 Species of Allium & authorities (incl. synonyms)	2 Fl. China	3 Middle Asia	4 Fl. USSR (+= Fl. As. med.)	5 Fl. Iran	6 Fl. Turkey	7 Fl. Europ.	8 Others	9 Fritsch & Friesen	10 B. Mathew	11 Accepted names for synonyms
jesdianum Vved.	—	—	210 Mol	—	—	—	—	—	—	=altissimum
" Boiss. et Buhse	—	146 Meg	—	119 Meg	—	—	—	Meg	—	see Kh & Fr 1994, Fritsch 1996
" " var. latipetalum Lipsky	—	—	—	—	—	—	—	Meg	—	=altissimum
" Boiss. ssp. angustitepalum (Wend.) Khassanov & R.M.Fritsch	—	—	—	—	—	—	—	Meg	—	see Kh & Fr 1994
jodanthum Vved.	—	27 Camp	+ Rhiz	—	—	—	—	Camp	—	
jubatum Macbride	—	—	—	—	112 All	100 All	—	—	57	
jucundum Vved.	—	51 Orei	61 Rhiz	—	—	—	—	Orei	—	
juldusicolum Regel	—	—	—	—	—	—	—	Ret	—	IK (Turkestan) 1879
junceum Jacq. ex Bak.	18 Rhiz	—	—	—	—	—	—	Ret	—	=przewalskianum
" Sm.	—	—	—	—	111 All	—	—	—	54	
" ssp. junceum	—	—	—	—	111 All	—	—	—	54a	
" " tridentatum Kollm., Özhatay et Koyuncu	—	—	—	—	111 All	—	—	—	54b	
kachrooi G.Singh	—	—	—	—	—	—	Rhiz	Ret	—	IK (Kashmir) 1977
kansuense Regel	35 Rhiz	—	—	—	—	—	—	Ret	—	China=sikkimense
karacae Koyuncu	—	—	—	—	—	—	Scor	—	—	IK (Turkey) 1994
karakense Regel	—	88 Cost	149 Porr	—	—	—	—	—	107	=filidens
karamanoglui Koyuncu et Kollm.	—	—	—	—	136 Mel	—	—	Mel	—	
karataviense Regel	—	200 Mini	195 Mol	s.n. Mel	s.n. Mel	—	—	Mini	—	See Kh & Fr 1994
karelinii P.Poljak	72a Schoe	76 Schoe	+ Rhiz	—	—	—	—	Schoe	—	Turk=imperfectly known China & +=schoenoprasum var. scaberrimum
karsianum Fom.	—	—	107 Hap	—	47 Cod	—	—	—	—	
karyeteinii Post [-ni]	—	—	—	—	116 All	—	—	—	102	
kaschianum Regel [kashch-]	58 Rhiz	47 Orei	57 Rhiz	—	—	—	—	Orei	—	
kastambulense Kollm.	—	—	—	—	61 Cod	—	—	—	—	
kastekii M.Pop.	—	54 Orei	+ Rhiz	—	—	—	—	Orei	—	
kaufmannii Regel	—	71 Ann	81 Rhiz	—	—	—	—	Ann	—	USSR=monadelphum
kazerouni Parsa	—	146 Meg	—	119 Meg	—	—	—	Meg	—	Asia & Iran=jesdianum
kermesinum Reichb. [kermesianum]	—	—	—	—	—	9 Rhiz	—	Orei	—	

kesselringii Regel	91 Hap	125 Caer	134 Hap	—	—	—	—	—	—	=schoenoprasoides
kharputense Freyn et Sint.	—	—	—	108 Mel	132 Mel	—	—	Mel	—	
khorasanicum Wend.	—	129 Kop	—	56 Scor	—	—	—	—	—	=kopetdagense
kingdonii Stearn	44 Rhiz	—	—	—	—	—	—	Ret	—	
kirindicum Bornm.	—	134 Brev	111 Hap	22 Scor	—	—	—	—	—	
klotzschi Regel	—	—	—	—	—	—	?	—	—	see Hook. rev. Stearn 1945
kochii Lange	—	—	—	—	—	95 All	—	—	85	BM=vineale
										Eur= " var. purpureum
koelzii (Wend.) Perss. et Wend.	—	—	—	Nect	—	—	—	Pseud	—	IK & Iran=Nectaroscordum koelzii
koenigianum Grossh.	—	—	—	—	6 Schoe	—	—	Scor?	—	see Friesen 1996
kokanicum Regel	—	53 Orei	63 Rhiz	—	—	—	—	Orei	—	
kollmannianum Brullo, Pavone et Salmeri	—	—	—	—	—	—	Brev	—	—	IK (Israel) 1991
komarovianum Vved.	84 Hap	—	4 Rhiz	—	—	—	—	Sacc	—	China=thunbergii
komarowii Lipsky [-ovii]	—	179 Comp	213 Mol	—	—	—	—	Comp	—	see Kh & Fr 1994
kondarinum Kamelin	—	27 Camp	—	—	—	—	—	—	—	nom. inval.;=jodanthum
kopetdagense Vved.	—	129 Kop	132 Hap	56 Scor	—	—	—	—	—	
korolkowii Regel [-ovii]	15 Rhiz	67 Orei	75 Rhiz	—	—	—	—	Orei	—	
" var. albidum Kuntze	—	—	112 Hap	—	—	—	—	—	—	=fibrosum
kossoricum Fom.	—	—	100 Hap	s.n. Cod	37 Scor	—	—	—	—	
" var. araraticum Miscz. ex Grossh.	—	—	99 Hap	—	—	—	—	—	—	=stamineum
kotschyi Boiss.	—	—	—	58 All	—	—	—	—	76	
krylovii K.Sobol.	—	—	—	—	—	—	Rhiz	Caes	—	IK (Tuva) 1949
										=mongolicum (see Friesen 1988)
kujukense Vved.	—	138 Vved	135 Hap	—	—	—	—	—	—	
kungii Nak.	51 Rhiz (p.p.)	—	—	—	—	—	—	Rhiz	—	=senescens
kunthianum Vved.	—	—	108 Hap	85 Cod	45 Cod	—	—	—	—	
kuramense Khassanov et Friesen	—	—	—	—	—	—	—	Ret	—	(Uzbekistan) Fritsch et al. 1998
kurdaicum Bajtenov	—	—	—	—	—	—	?	—	—	IK (Kazakhstan) 1995; RMF
										doubtful if good sp.
kurrat Schweinf. ex Krause (cult.)	—	—	—	—	—	—	7 All* Pal s.n. All	—	1b	*=ampeloprasum
kurssanovii M.Pop.	—	62 Orei	+ Rhiz	—	—	—	—	Orei	—	
kurtzianum Asch. et Sint. ex Kollm.	—	—	—	—	54 Cod	—	—	—	—	
kuschakewiczii Regel	—	116 Avu	121 Hap	—	—	—	—	—	—	=griffithianum
kysylkumii Kamelin	—	34 Camp	—	—	—	—	Rhiz	Camp	—	IK (Middle Asia) 1980

Column 1 Species of Allium & authorities (incl. synonyms)	2 Fl. China	3 Middle Asia	4 Fl. USSR (+= Fl. As. med.)	5 Fl. Iran	6 Fl. Turkey	7 Fl. Europ.	8 Others	9 Fritsch & Friesen	10 B. Math-ew	11 Accepted names for synonyms
laceratum Boiss. et Noë	—	—	—	90 Mol	14 Mol	—	—	—	—	Iran=eriophyllum var. laceratum Turk=longisepalum var. "
laceratum Freyn	—	—	98 Hap	—	21 Brev	—	—	—	—	USSR=lacerum; Turk=callidictyon
lacerum Freyn	—	—	98 Hap	21 Scor	21 Brev	—	—	—	—	Iran & Turk=callidictyon
" var. ochroleucum Freyn et Sint.	—	—	—	—	21 Brev	—	—	—	—	=callidictyon
lachnophyllum Paine [ex Dinsm.]	—	—	—	—	—	—	30 Mel*	Mel	—	* possibly = form of orientale
lacteum Sm.	—	—	—	—	12 Mol	—	23 Mol*	—	—	=neapolitanum
laeve Wend. et Bothmer	—	—	—	80 All	85 All	—	—	—	14	BM=macrochaetum Turk= " ssp. macrochaetum
lagarophyllum Brullo, Pavone et Tzan.	—	—	—	—	—	—	Scor	—	—	IK (Greece) 1993
lalesaricum Freyn et Bornm.	—	—	—	31 Scor	—	—	—	—	—	
lallemantii Regel et Rach.	—	—	—	—	—	—	—	Mel	—	=tulipifolium
lamondiae Wend.	—	—	—	39 Scor	—	—	—	—	—	
lancifolium Stearn	10a Brom	—	—	—	—	—	—	—	—	=wallichii var. platyphyllum
lancipetalum Y.P.Hsu	—	—	—	—	—	—	Hap	—	—	IK (China) 1987
lasiophyllum Vved.	—	122 Caer	124 Hap	—	—	—	—	—	—	
latei Aitch. et Bak.										see yatei
latifolium Jaub. et Spach.	—	—	190 Mol	98 Acan	120 Acan	—	—	Acan	—	=akaka, RMF=?akaka
" Gilib.	—	—	2 Oph	—	—	—	—	—	—	=ursinum
latissimum Prokh.	1 Ang	—	1 Ang	—	—	—	—	Ang	—	China & USSR=victorialis, NWF=ochotense
laxum Don	—	—	37 Rhiz	—	—	—	—	Rhiz	—	=angulosum
ledebourianum Roem. et Schult.	73 Schoe	—	83 Rhiz	—	—	—	Ohwi	Schoe	—	Ohwi= schoenoprasum var. foliosum
" var. intermedium Kryl.	—	—	—	—	—	—	Schoe	Schoe	—	=altyncolicum (Friesen 1996)
ledschanense Conrath et Freyn	—	—	154 Porr	—	109 All	—	—	—	58	=?aucheri
lefkarense Brullo, Pavone et Salmeri	—	—	—	—	—	—	Cod	—	—	IK (Cyprus) 1993
lehmanii Lojac. [-mannii]	—	—	—	—	—	57c Cod	—	—	—	=pallens ssp. siciliense
lehmannianum Merckl.	—	98 Mult	144 Porr	—	—	—	—	—	77	
" var. bungei Boiss.	—	100 Mult	—	59 All	—	—	—	—	100	=borszczowii
" " kokanicum Regel	—	99 Mult	143 Porr	—	—	—	—	—	79	=ferganicum
" auct.	—	—	+ Porr	—	—	—	—	—	—	=borszczowii

lenkoranicum Miscz. ex Grossh. [lencoranicum]	—	136 Cod	106 Hap	84 Cod	—	—	—	—	—	
leonidis Grossh.	—	—	—	—	—	—	Mel	Meg	—	Kud 1992; USSR
lepidum Kunth p.p.	—	—	108 Hap	85 Cod	45 Cod	—	—	—	—	=kunthianum
" Ledeb.	88 Hap	121 Caer	127 Hap	—	—	—	—	—	—	=pallasii
leptomorphum Vved. ex Kaschtsch. et E.Nikit.	—	63 Orei	+ Rhiz	—	—	—	—	Orei	—	
leptophyllum Schur	—	—	—	—	—	—	BM	Rhiz	—	BM=senescens, NWF=lusitanicum
leucanthum C.Koch	—	—	174 Porr	82 All	—	—	7 All*	—	6	*=ampeloprasum Iran= " ssp. ampeloprasum
leucocephalum Turcz.	20 Rhiz	—	9 Rhiz	—	—	—	—	Ret	—	
" auct. non Turcz.	—	—	—	—	—	—	Rhiz	—	—	=schischkinii (Friesen 1988)
leucosphaerum Aitch. et Bak.	—	114 Avu	112 Hap	38 Scor	—	—	—	—	—	=fibrosum
liangshanense Z.Y.Zhu	—	—	—	—	—	—	Brom	—	—	IK (China) 1991
libani Boiss.	—	—	—	—	—	—	30 Mel*	Acan	—	* may be form of orientale
lilacinum Royle ex Regel	—	—	—	14 Rhiz	—	—	—	—	—	=roylei
lineare L.	22 Rhiz	1 Ret	13 Rhiz	—	—	11 Rhiz	—	Ret	—	
" var. maackii Maxim.	—	—	11 Rhiz	—	—	—	—	—	—	=maackii
" " strictum Kryl.	23 Rhiz	—	14 Rhiz	—	—	—	—	—	—	=strictum Schrad.
" Miller	—	—	—	—	—	—	—	—	1	=ampeloprasum
" Ten.	—	—	—	—	—	—	—	—	80a	=?guttatum ssp. sardoum
lipskyanum Vved.	—	185 Reg	+ Mol	—	—	—	—	Reg	—	see Kh & Fr 1994
listera Stearn [listeria]	1a Ang	—	—	—	—	—	—	Ang	—	China=victorialis var. listeria
litvinovii Drob. ex Vved.	—	119 Caer	+ Hap	—	—	—	—	—	—	
lojaconoi Brullo, Lanfranco et Pavone	—	—	—	—	—	—	Brev	—	—	IK (Malta) 1982
longanum Pamp.	—	—	—	—	—	29 Mol	19 Mol*	—	—	
longicollum Wend.	—	—	—	64 All	—	—	—	—	49	
longicuspis Regel	—	83 All	162 Porr	72 All	75 All	—	—	—	19	
longiradiatum (Regel) Vved.	—	28 Camp	29 Rhiz	—	—	—	—	Camp	—	
longisepalum Bertol.	—	—	—	—	14 Mol	—	—	—	—	
" var. laceratum (Boiss. et Noë) Wend.	—	—	—	—	14 Mol	—	—	—	—	
longispathum Delaroche	—	—	—	—	38 Cod	—	Pal 9 Cod	—	—	=paniculatum ssp. paniculatum
" Redouté	—	—	—	—	—	56a Cod	14 Cod*	—	—	= " or p. " "
" ssp. rupestre (Steven) Nyman	—	—	—	—	—	59 Cod	—	—	—	=rupestre
longistamineum Royle	—	—	—	—	—	—	Rhiz	—	—	=stracheyi (Hook. rev. Stearn 1945)

Column 1 Species of Allium & authorities (incl. synonyms)	2 Fl. China	3 Middle Asia	4 Fl. USSR (+= Fl. As. med.)	5 Fl. Iran	6 Fl. Turkey	7 Fl. Europ.	8 Others	9 Fritsch & Friesen	10 B. Mathew	11 Accepted names for synonyms
longistylum Baker	62 Rhiz	—	—	—	—	—	—	Orei	—	
longivaginatum Wend.	—	—	—	30 Scor	—	—	—	—	—	longe- in IK but longi- correct
lopadusanum Bartolo, Brullo et Pavone	—	—	—	—	—	—	Brev	—	—	IK (Sicily) 1986
loratum Baker	—	—	—	—	—	—	26 Mol	Pseud	—	Hook. rev. Stearn 1945
loscosii Richter	—	—	—	—	—	90 All	—	—	62a	Eur=variant of sphaerocephalon; BM=s. ssp. sphaerocephalon
lownei Baker	—	—	—	—	—	—	Pal 8 Mol	—	—	=carmeli var. roseum
lucens E.Nikit.	—	176 Comp	+ Mol	—	—	—	—	Comp	—	+=elatum; Asia & RMF, nom. inval.,=macleanii
lusitanicum Lam.	—	—	—	—	—	3 Rhiz	—	Rhiz	—	Eur=senescens ssp. montanum; see also Stearn 1978
luteolum Halácsy	—	—	—	—	—	67 Cod	—	—	—	
lutescens Vved.	—	25 Camp	28 Rhiz	—	—	—	—	Camp	—	
lycaonicum Siehe ex Hayek	—	—	—	—	137 Mel	—	—	Mel	—	RMF=?chrysantherum
maackii Prokh. ex Komarov	—	—	11 Rhiz	—	—	—	—	Ret	—	
macedonicum Zahar.	—	—	—	—	—	61 Cod	—	—	—	
macleanii Baker	—	176 Comp	—	128 Meg	—	—	—	Comp	—	
macranthum Baker	12 Brom	—	—	—	—	—	—	—	—	
macrochaetum Boiss. et Hausskn.	—	—	—	—	85 All	—	—	—	14	
" ssp. macrochaetum	—	—	—	—	85 All	—	—	—	—	
" " tuncelianum Kollm.	—	—	—	—	85 All	—	—	—	20	BM=tuncelianum
macror[r]hizon Regel [-um]	—	55 Orei	66 Rhiz	—	—	—	—	Orei	—	=tianschanicum
macrorrhizum Boiss. [-orh-]	57 Rhiz	46 Orei	56 Rhiz	12 Rhiz	—	—	—	—	—	=hymenorrhizum
macrostemon Bunge [-um]	89 Hap	—	128 Hap	—	—	—	—	—	—	
magicum auct., an L.	—	—	—	—	122 Mel	106 Mel	29b Mel* Pal 24 Mel	Mel	—	=nigrum, RMF=?nigrum *=n. ssp. nigrum; see Seisums 1998
mairei Lévl.	42 Rhiz	—	—	—	—	—	Col	—	—	Hanelt et al. 1992
majale Ten.	—	—	—	—	15 Mol	—	25 Mol*	—	—	=roseum
majus Vved.	—	178 Comp	+ Mol	—	—	—	—	Comp	—	see Kh & Fr 1994
makmelianum Post	—	—	—	—	—	—	—	—	75	Lebanon, Syria
malyschevii Friesen	—	—	—	—	—	—	Rhiz	Ret	—	IK (S.Siberia; Mongolia) 1987

maniaticum Brullo et Tzan.	—	—	—	—	—	—	Brev	—	—	IK (Greece) 1989
maowenense J.M.Xu	—	—	—	—	—	—	Hap	—	—	IK (China) 1994
marathasicum Brullo, Pavone et Salmeri	—	—	—	—	—	—	Cod	—	—	IK (Cyprus) 1993
mareoticum Bornm. et Gauba	—	—	—	—	—	—	3 All*	—	78	
margaritaceum auct.	—	—	149 Porr	—	—	—	—	—	—	=filidens
" Smith	—	—	—	—	106 All	97b All	2 All*	—	80b	*guttatum; Eur, Turk & BM=g. ssp. sardoum or ssp. guttatum
" ssp. dalmaticum A.Kerner ex Janchen	—	—	—	—	—	—	—	—	80c	=guttatum ssp. dalmaticum
" " guttatum (Steven) Nyman	—	—	—	—	—	97 All	—	—	—	=guttatum ssp. guttatum
" " tenorii [(Parl.)] Richter		—	—	—	—	—	2 All*	—	80b	*=guttatum; BM=g. ssp. sardoum
" var. affine (Ledeb.) Regel	—	—	156 Porr	—	107 All	—	—	—	82	=affine
" " battandieri Maire et Weiller	—	—	—	—	—	—	2 All*	—	80b	*=guttatum ssp. margaritaceum; BM=g. ssp. sardoum
" " bulbiferum Batt. et Trab.	—	—	—	—	—	—	1 All*	—	85	=vineale
" " compactum Batt. et Trab.	—	—	—	—	—	—	1 All*	—	85	= "
" " faurei Maire	—	—	—	—	—	—	2 All*	—	80b	*=guttatum; BM=g. ssp. sardoum
" " gorumsense Regel	—	—	—	—	—	—	—	—	83	=gorumsense
" " guttatum (Steven) Gay	—	—	155 Porr	—	106 All	97a All	2 All*	—	80a	* & USSR=guttatum; Turk, Eur & BM=g. ssp. guttatum
" " papillosum Lindberg	—	—	—	—	—	—	9 All*	—	48	=baeticum
" " robustum Maire	—	—	—	—	—	—	4a All*	—	62a	=sphaerocephalon ssp. sphaerocephalon
" " rubellum Boiss.	—	—	—	—	106 All	—	—	—	80c	=?guttatum ssp. dalmaticum
" " scabrum Regel	—	—	156 Porr	—	107 All	—	—	—	82	=affine
" " tenorii Parl.	—	—	—	—	—	—	2 All*	—	80b	*=guttatum; BM=g. ssp. sardoum
" " typicum Regel	—	—	—	—	—	—	2 All*	—	80b	" " " " "
margaritae B.Fedtsch.	—	133 Brev	94 Hap + Porr	—	—	—	—	—	—	
margaritiferum Vved.	—	91 Cost		—	—	—	—	—	109	
marginatum Janka	—	—	—	—	—	—	Cod	—	—	=paniculatum ssp. fuscum (Fl. Roman.)
mariae E.Bordz.	—	—	196 Mol	—	—	—	—	Pseud	—	
maritimum Raf.	—	—	—	—	?All	51 Scor	—	—	—	Eur=obtusiflorum Turk prob.=sphaerocephalon
marschallianum Vved. [-l-]	—	—	70 Rhiz	—	—	4 Rhiz	—	Orei	—	Eur=saxatile, NWF=?saxatile
massaessylum Batt. et Trab.	—	—	—	—	—	25 Mol	24 Mol*	—	—	
materculae Bordz.	—	—	191 Mol	100 Acan	—	—	—	Acan	—	

Column 1 Species of Allium & authorities (incl. synonyms)	2 Fl. China	3 Middle Asia	4 Fl. USSR (+= Fl. As. med.)	5 Fl. Iran	6 Fl. Turkey	7 Fl. Europ.	8 Others	9 Fritsch & Friesen	10 B. Math-ew	11 Accepted names for synonyms
maximowiczii Regel	73 Schoe	—	84 Rhiz	—	—	—	Ohwi	Schoe	—	China=ledebourianum; NWF=good sp.
" var. shibutsuense (Kitam.) Ohwi	—	—	—	—	—	—	Ohwi	—	—	
maximowiczii var. yezomonticola (Hara) T.Shimizu	—	—	—	—	—	—	Schoe	Schoe	—	IK (Japan) 1983 NWF=maximowiczii
megalobulbon Regel	—	—	—	—	—	—	—	Orei	—	Middle Asia
melanantherum Pančić	—	—	—	—	—	69 Cod	—	—	—	
melananthum Coincy [melanthum]	—	—	—	—	—	92 All	—	—	70	
meliophilum Juz.	—	—	—	—	Nect	Nect	—	—	—	=Nectaroscordum siculum ssp. bulgaricum
melitense (Somm. et Car.-Gatto) Ciferri et Giacom.	—	—	—	—	—	—	All	—	—	=ampeloprasum var. melitense Somm. et Car.-Gatto (IK)
meteoricum Heldr. et Hausskn. ex Halácsy	—	—	—	—	—	44 Scor	—	—	—	
micranthum Wend.	—	—	—	53 Scor	—	—	—	Brevid	—	see Fritsch et al. 1998
microbulbum Prokh.	—	—	86 Phyll	—	—	—	—	Cep	—	NWF=altaicum
microdictyon Prokh. [-um]	1 Ang	—	1 Ang	—	—	—	—	Ang	—	China & USSR=victorialis
microspathum Ekberg	—	—	—	—	26 Scor	—	—	—	—	
minoricense Llorens	—	—	—	—	—	—	?	—	—	nom. illeg. (Garbari et al. 1991)
minutiflorum Regel	—	—	—	101 Acan	—	—	—	Acan	—	
minutum Vved.	—	131 Minu	119 Hap	—	—	—	—	—	—	
mirum Wend.	—	—	—	133 Thaum	—	—	—	Thaum	—	
mirzajevii Tscholok.	—	—	—	—	—	—	Rhiz	Orei	—	=gunibicum (Kud 1992)
miserabile Wend.	—	—	—	51 Scor	—	—	—	Brevid	—	see Fritsch et al. 1998
mishtshenkoanum Grossh. [mishtsch-]	—	—	156 Porr	—	107 All	—	—	—	82	USSR & Turk=affine, BM=?affine
modestum Boiss.	—	—	—	—	—	—	14 Cod* Pal 11 Cod 7 All*	—	—	Pal=desertorum; *=paniculatum
mogadorense Willd. [ex Schult. et Schult.f.]	—	—	—	—	—	—		—	1	=ampeloprasum
mogoltavicum Vved.	—	173 Acmo	+ Mol	—	—	—	—	Acmo	—	=taeniopetalum ssp. mogoltavicum (Fritsch et al. 1998)

moly L.	—	—	—	—	—	33 Mol	—	—	—	
" ssp. massaessylum (Batt. et Trab.) Vindt	—	—	—	—	—	25 Mol	24 Mol*	—	—	=massaessylum
moly var. stramineum (Boiss. et Reut.) Wolley-Dod	—	—	—	—	—	—	34 Mol	—	—	=scorzonerifolium var. xericiense (Stearn 1978)
" " xericiense Perez-Lara	—	—	—	—	—	—	34 Mol	—	—	= " " " " "
monadelphum Turcz. ex Kar. et Kir.	71 Schoe	—	—	—	—	—	—	Ann	—	China=atrosanguineum
" Less. ex Kunth	—	—	81 Rhiz	—	—	—	—	—	—	
" var. atrosanguineum Regel	—	70 Ann	—	—	—	—	—	—	—	=atrosanguineum
" " bucharicum Regel	—	72 Ann	—	—	—	—	—	—	—	=fedschenkoanum
" " fed[t]schenkoanum Regel	—	72 Ann	—	—	—	—	—	—	—	= "
" " humile Regel	—	72 Ann	—	—	—	—	—	—	—	= "
" " kaufmannii Regel	—	71 Ann	—	—	—	—	—	—	—	=kaufmannii
monanthum Maxim.	92 Hap	—	177 Mol	—	—	—	Ohwi	—	—	Micr; Hanelt et al. 1992
mongolicum Regel	29 Rhiz	—	—	—	—	—	—	Caes	—	
monophyllum Vved. ex Czerniak	—	198 Acan	187 Mol	94 Acan	—	—	—	Acan	—	
monspessulanum Willd.	—	—	—	—	—	—	Cod	—	—	=dentiferum (BPS 1991)
" Gouan	—	—	—	—	122 Mel	—	29b Mel*	Mel	—	=nigrum, *=n. ssp. nigrum
" ssp. pruinatum (Link) Richter	—	—	—	—	—	—	—	—	69	=pruinatum
montanum Smith	—	—	—	—	49 Cod	—	14 Cod*	—	—	Turk=sibthorpianum; *=paniculatum
" var. subunivalve Ten.	—	—	—	—	—	—	15 Cod*	—	—	=cupanii
" Schmidt	51 Rhiz	—	44 Rhiz	—	—	3 Rhiz	—	Rhiz	—	China & USSR=senescens, Eur=s. ssp. montanum, NWF=lusitanicum
" " ssp. leptophyllum (Schur) Soó	—	—	—	—	—	—	Rhiz (BM)	Rhiz	—	BM=senescens, NWF=lusitanicum
" Schrank	—	—	—	—	—	18 Schoe	—	—	—	=schoenoprasum
montibaicalense Friesen	—	—	—	—	—	—	—	Ret	—	IK (Siberia) 1992
morrisonense Hay.	84 Hap	—	—	—	—	—	—	Sacc	—	=thunbergii
moschatum L.	—	—	96 Hap	—	30 Scor	40 Scor	—	—	—	
" var. dubium Regel	15 Rhiz	67 Orei	75 Rhiz	—	—	—	—	Orei	—	=korolkowii
" " brevipedunculatum Regel	15 Rhiz	67 Orei	75 Rhiz	—	—	—	—	—	—	= "
" d'Urv.	—	—	—	—	20 Brev	—	—	—	—	=cupanii ssp. hirtovaginatum
" auct. p.p.	—	—	94 Hap	—	—	—	—	—	—	=margaritae
" auct. p.p.	—	—	95 Hap	—	—	—	—	—	—	=inaequale
motor Kamelin et Levichev	—	165 Acmo	—	—	—	—	—	Acmo	—	See Kh & Fr 1994
multibulbosum Jacq.	—	—	—	—	122 Mel	106 Mel	29a Mel*	Mel	—	=nigrum, *=n. ssp. multibulbosum

Column 1 Species of Allium & authorities (incl. synonyms)	2 Fl. China	3 Middle Asia	4 Fl. USSR (+= Fl. As. med.)	5 Fl. Iran	6 Fl. Turkey	7 Fl. Europ.	8 Others	9 Fritsch & Friesen	10 B. Math-ew	11 Accepted names for synonyms
multiflorum Desf.	—	—	—	—	—	—	7 All*	—	1	=ampeloprasum
" auct. non Desf.	—	—	—	—	—	73 Cod	—	—	—	=hymettium
" DC.	—	—	—	—	—	—	—	—	10	=polyanthum
multiflorum var. violaceopurpureum C.Koch	—	—	—	—	s.n. All	—	—	—	101	Turk=imperfectly known BM=?dictyoprasum
multitabulatum S.Cic.	—	—	—	—	—	—	—	Cep	—	=proliferum
myrianthum Boiss.	—	—	—	87 Cod	69 Cod	71 Cod	14 Cod*	—	—	Eur perhaps=convallarioides *=paniculatum
nabelekii Kamelin et Seisums	—	—	—	—	—	—	—	Pseud	—	IK (Turkey) 1996; RMF=cardiostemon (pink form)
nanodes Airy-Shaw	5 Ang	—	—	—	—	—	—	Ang	—	
narcissiflorum Vill.	—	—	—	—	—	15 Rhiz	Nark	—	—	Hanelt et al. 1992
" ssp. insubricum (Boiss. et Reut.) Ciferri	—	—	—	—	—	16 Rhiz	Nark	—	—	=insubricum Hanelt et al. 1992
" var. insubricum (Boiss. et Reut.) Fiori	—	—	—	—	—	—	16 Rhiz Nark	—	—	=insubricum (Stearn 1978); Nark (Hanelt et al. 1992)
neapolitanum Cirillo	—	—	—	—	12 Mol	28 Mol	23 Mol* Pal 1 Mol	—	—	
" ssp. philippi Radić	—	—	—	—	—	—	Mol	—	—	Radić 1989
" var. angustifolium Täckh. et Drar	—	—	—	—	—	—	23 Mol*	—	—	=neapolitanum
" " breviradium Halácsy	—	—	—	—	—	27 Mol	—	—	—	=breviradium
nebrodense Guss.	—	—	—	—	—	68a Cod	—	—	—	=flavum ssp. flavum
negevense Kollm.	—	—	—	—	—	—	Pal 6 Mol	—	—	
negrianum Maire et Weiller	—	—	—	—	—	—	2 All*	—	80b	*=guttatum; BM=g. ssp. sardoum
nemrutdaghense Kit Tan et F.Sorger	—	—	—	—	133a Mel	—	Mel	Mel	—	Turk vol. 10
nereidum Hance	89 Hap	—	—	—	—	—	—	—	—	=macrostemon
neriniflorum (Herb.) Baker	99 Cal	—	—	—	—	—	—	Cal	—	=Caloscordum neriniflorum Herb. (BM); RMF=good sp.
nerinifolium Baker	—	—	226 Cal	—	—	—	—	—	—	mistake for neriniflorum (BM)
nevsehirense Koyuncu et Kollm.	—	—	—	—	102 All	—	—	—	59	
nevskianum Vved. [ex Wend.]	—	196 Kal	+ Mol	93 Acan	—	—	—	Kal	—	see Kh & Fr 1994
nigritanum A.Chev.	—	—	—	—	—	—	10 Schoe*	Cep	—	=cepa

nigrum auct.	—	—	—	—	—	—	29a Mel*	—	—	=nigrum ssp. multibulbosum
" L.	—	—	—	—	122 Mel	106 Mel	29, 29b Mel* Pal 24 Mel	Mel	—	*=nigrum & ssp. nigrum; see Seisums 1998
" ssp. multibulbosum (Jacq.) Holmboe	—	—	—	—	—	—	29a Mel*	—	—	
" " nigrum	—	—	—	—	—	—	29b Mel*	—	—	
" var. atropurpureum (Waldst. et Kit.) Vis.	—	—	—	—	123 Mel	105 Mel	—	Mel	—	=atropurpureum
" " bulbiferum Grenier et Godron	—	—	—	—	—	—	29b Mel*	—	—	=nigrum ssp. nigrum
" " cyrilli (Ten.) Fiori	—	—	—	—	124 Mel	—	29a Mel*	Mel	—	Turk and RMF=cyrilli; *=nigrum ssp. multibulbosum
" " denticulatum Pamp.	—	—	—	—	—	—	29b Mel*	—	—	=nigrum ssp. nigrum
" " dumetorum (Feinbr. et Szelubsky) Mout.	—	—	—	—	—	—	Pal 24 Mel	—	—	Pal=nigrum; RMF=dumetorum
" " multibulbosum Rouy	—	—	—	—	—	—	29a Mel*	—	—	=nigrum ssp. multibulbosum
" " papillosum Pamp.	—	—	—	—	—	—	30 Mel*	—	—	=orientale
nipponicum Franch. et Sav.	89 Hap	—	—	—	—	—	Ohwi	—	—	Ohwi=grayi; Chin=macrostemon
nitens Sauzé et Maillard	—	—	—	—	—	—	—	—	85	=vineale
niveum Roth	—	—	—	—	8 Mol	—	18a Mol*	—	—	=subhirsutum, *=s. ssp subhirsutum
noëanum Reut. ex Regel	—	—	—	106 Mel	135 Mel	—	—	Mel	—	
notabile Feinbr.	—	—	—	75 All	—	—	—	—	34	
nothum Vved.	—	87 All	142 Hap	—	—	—	—	—	115	=turkestanicum (BM section ?Cod)
nuristanicum Kitamura	—	—	—	4 Rhiz	—	—	—	Ret	—	
nutans L.	52 Rhiz	38 Rhiz	45 Rhiz	—	—	—	—	Rhiz	—	
obliquum L.	56 Rhiz	68 Pet	53 Rhiz	—	—	14 Rhiz	—	Pet	—	
obtusiflorum Requien ex Grenier et Godron	—	—	—	—	—	—	25 Mol*	—	—	=roseum
" DC.	—	—	—	—	—	51 Scor	14 Cod*	Scor	—	see BPST 1994; *=paniculatum
obtusifolium Klotzsch	54 Rhiz	45 Orei	—	11 Rhiz	—	—	—	Orei	—	=carolinianum
ochotense Prokh. [okho-]	1 Ang	—	1 Ang	—	—	—	—	Ang	—	China & USSR=victorialis
ochroleucum Waldst. et Kit.	—	—	—	—	—	8 Rhiz	—	Orei	—	Eur=ericetorum
odoratissimum Desf.	—	—	—	—	—	—	25 Mol*	—	—	=roseum
odorum L.	26 Rhiz	—	36 Rhiz	—	—	—	—	But	—	China & NWF=ramosum
" auct.	—	—	+ Rhiz	—	—	—	—	—	—	=chinense
" auct. hisp.	—	—	—	—	—	Noth	—	—	—	=Nothoscordum inodorum
" auct. japon.	—	—	—	—	—	—	Ohwi	—	—	=tuberosum
okhotense Prokh.										see ochotense

Column 1 Species of Allium & authorities (incl. synonyms)	2 Fl. China	3 Middle Asia	4 Fl. USSR (+= Fl. As. med.)	5 Fl. Iran	6 Fl. Turkey	7 Fl. Europ.	8 Others	9 Fritsch & Friesen	10 B. Math-ew	11 Accepted names for synonyms
oleraceum L.	—	—	104 Hap	—	s.n. Cod	63 Cod	—	—	—	Turk=imperfectly known
" ssp. girerdii J.-M.Tison	—	—	—	—	—	—	Cod	—	—	IK (France) 1993
" var. complanatum Fries	—	—	—	—	—	—	63 Cod	—	—	=oleraceum (Stearn 1978)
oliganthum Kar. et Kir.	—	74 Schoe	85 Rhiz	—	—	—	—	Schoe	—	
" var. elongatum Kar. et Kir.	15 Rhiz	67 Orei	75 Rhiz	—	—	—	—	—	—	=korolkowii
" auct.	—	—	75 Rhiz	—	—	—	—	—	—	=korolkowii
olivieri Boiss. [oliveri]	—	—	—	107 Mel	—	—	—	Mel	—	
oltense Grossh.	—	—	—	—	81 All	—	—	—	22	
olympicum Boiss.	—	—	—	—	60 Cod	—	—	—	—	
omeiense Z.Y.Zhu	—	—	—	—	—	—	Brom	—	—	IK (China) 1989
omiostema Airy-Shaw	31 Rhiz	—	48 Rhiz	—	—	—	—	Caes	—	=bidentatum
opacum Rech. f.	—	—	—	—	40 Cod	—	—	—	—	
ophiophyllum Vved.	—	103 Scor	122 Hap	—	—	—	—	—	—	
ophiopogon Lévl.	84 Hap	—	—	—	—	—	—	Sacc	—	=thunbergii
ophioscorodon G.Don [Link]	—	—	—	—	—	75 All	6b All*	—	—	=sativum var. ophioscorodon
oporinanthum Brullo, Pavone et Salmeri	—	—	—	—	—	—	Cod	—	—	IK (Spain, France) 1997
oreodictyum Vved.	—	10 Ret	20 Rhiz	—	—	—	—	Ret	—	
oreophiloides Regel	—	128 Caer	133 Hap	25 Scor	—	—	—	—	—	
" ssp. oreophiloides	—	—	—	25 Scor	—	—	—	—	—	
" " salangense Wend.	—	—	—	25 Scor	—	—	—	—	—	
oreophilum C.A.Mey.	—	205 Porph	180 Mol	92 Porph	19 Porph	—	—	Porph	—	
oreoprasoides Vved.	—	9 Ret	19 Rhiz	—	—	—	—	Ret	—	
oreoprasum Schrenk	24 Rhiz	8 Ret	35 Rhiz	—	—	—	—	Ret	—	
oreoscordum Vved.	—	18 Ret	6 Rhiz	—	—	—	—	Ret	—	
orientale Boiss.	—	—	—	—	131 Mel	109 Mel	30 Mel* Pal 27 Mel	Mel	—	
oschaninii O.Fedtsch. [-nini]	—	80 Cep	92 Cep	19 Cep	Cep	—	—	Cep	—	see OF 1906
ostrowskianum Regel [ostrov-]	—	205 Porph	180 Mol	92 Porph	19 Porph	—	—	Porph	—	=oreophilum
otschiauriae Tscholok.	—	—	—	—	—	—	Rhiz	Orei	—	IK (Caucasus) 1965; see Kud 1992 =albovianum
ousensanense Nakai	89 Hap	—	—	—	—	—	—	—	—	=macrostemon
ovalifolium Hand.-Mazz.	2 Ang	—	—	—	—	—	—	Ang	—	

" var. cordifolium (J.M.Xu) J.M.Xu	—	—	—	—	—	—	Ang	—	—	IK (China) 1991
" " leuconeurum J.M.Xu	2a Ang	—	—	—	—	—	—	—	—	
oviflorum Regel	12 Brom	—	—	—	—	—	—	—	—	=macranthum
paczoskianum Tuzson	—	—	102 Hap	—	—	68b Cod	—	—	—	=carinatum ssp. pulchellum (Stearn 1978) (USSR=pulchellum Eur=flavum ssp. tauricum)
paepalanthoides Airy-Shaw	40 Rhiz	—	—	—	—	—	—	Sacc	—	
palaestinum Kollm.	—	—	—	—	—	—	Pal 1 Mol	—	—	variant of neapolitanum
palentinum Losa et Montserrat	—	—	—	—	—	12 Rhiz	—	Orei	—	
pallasii Murr.	88 Hap	121 Caer	127 Hap	—	—	—	—	—	—	
" var. verticillatum Regel	—	—	185 Mol	—	—	—	—	Vert	—	=verticillatum
pallens L.	—	—	—	—	41 Cod	57 Cod	14 Cod* Pal 10 Cod	—	—	*=paniculatum
" ssp. christophori Radić	—	—	—	—	—	—	Cod	—	—	Radić 1989
" " pallens	—	—	—	—	41 Cod	57a Cod	—	—	—	
" " siciliense Stearn	—	—	—	—	—	57c Cod	—	—	—	
" " tenuiflorum (Ten.) Stearn	—	—	—	—	—	57b Cod	—	—	—	
" var. coppoleri (Tineo) Parl.	—	—	—	—	41 Cod	—	14 Cod* Pal 10 Cod	—	—	Turk=pallens ssp. pallens; Pal=pallens; *=paniculatum
" " savii (Parl.) Cesati, Passer. et Gibelli	—	—	—	—	—	—	Cod	—	—	=savii (BPSS 1994)
" Vved.	—	—	110 Hap	—	—	—	—	—	—	=convallarioides
pamiricum Wend.	—	115 Avu	—	46 Scor	—	—	—	—	—	
pangasicum I.Turakulov	—	163 Acmo	—	—	—	—	—	Acmo	—	IK (Tianshan) 1986; see Kh & Fr 1994
paniculatum auct.	—	—	+Hap	—	38 Cod	—	—	—	—	+=lenkoranicum & praescissum Turk=paniculatum ssp. fuscum
" L.	—	—	105 Hap	83 Cod	38 Cod	56 Cod	14 Cod* Pal 9 Cod	—	—	
" ssp. euboicum (Rech.f.) Stearn	—	—	—	—	—	56c Cod	—	—	—	
" " fuscum (Waldst. et Kit.) Arcangeli	—	—	—	—	38 Cod	56b Cod	Pal 9 Cod	—	—	
" " jacobi Radić	—	—	—	—	—	—	Cod	—	—	Radić 1989
" " pallens (L.) K.Richter	—	—	—	—	—	—	57a Cod	—	—	=pallens ssp. pallens (Stearn 1978)
" " paniculatum	—	—	—	—	38 Cod	56a Cod	Pal 9 Cod	—	—	
" " rupestre (Steven) K.Richter	—	—	—	—	—	—	59 Cod	—	—	=rupestre (Stearn 1978)

Column 1 Species of Allium & authorities (incl. synonyms)	2 Fl. China	3 Middle Asia	4 Fl. USSR (+= Fl. As. med.)	5 Fl. Iran	6 Fl. Turkey	7 Fl. Europ.	8 Others	9 Fritsch & Friesen	10 B. Math-ew	11 Accepted names for synonyms
paniculatum ssp.salinum (Debeaux) Botté et Kerguélen	—	—	—	—	—	—	Cod	—	—	=savii (BPSS 1994)
" " stearnii (Pastor et Valdés) O. de Bolos, R.M. Masalles et J.Vigo	—	—	—	—	—	—	Cod	—	—	=stearnii (IK 1987)
" " tenuiflorum (Ten.) Brand	—	—	—	—	—	57b Cod	—	—	—	=pallens ssp. tenuiflorum
" " villosulum (Hal.) Stearn	—	—	—	—	38 Cod	56d Cod	—	—	—	
" var. brevicaule (Boiss. et Bal.) Regel	—	—	—	—	51 Cod	—	—	—	—	=brevicaule
" " legitimum Ledeb.	—	135 Cod	105 Hap	—	—	—	—	—	—	Asia=praescissum; USSR=paniculatum
" " longispathum (Delaroche) Regel	—	—	—	—	38 Cod	—	14 Cod* Pal 9 Cod	—	—	=paniculatum ssp. paniculatum
" " macilentum Ledeb. p.p.	—	—	109 Hap	—	43 Cod	—	—	—	—	=rupestre
" " " " " p.p.	—	—	108 Hap	—	45 Cod	—	—	—	—	=kunthianum
" " pallens (L.) Regel or Gren et Godron	—	—	—	—	41 Cod	57a Cod	14 Cod* Pal 10 Cod	—	—	=pallens ssp. pallens (*=paniculatum)
" " " Boiss. [p.p.]	—	—	109 Hap	—	43 Cod	—	—	—	—	=rupestre
" " rhodopeum (Velen.) Stoj. et Stef.	—	—	—	—	38 Cod	—	—	—	—	=paniculatum ssp. villosulum
" " rupestre (Steven) Regel	—	—	109 Hap	—	43 Cod	—	—	—	—	=rupestre
" " salinum Debeaux	—	—	—	—	—	—	Cod	—	—	=savii (BPSS 1994)
" " tenuiflorum (Ten.) Regel	—	—	—	—	—	—	57b Cod	—	—	=pallens ssp. tenuiflorum (Stearn 1978)
" " typicum Regel	—	—	—	—	38 Cod	—	14 Cod*	—	—	Turk=paniculatum ssp. paniculatum, *=paniculatum
" " villosulum Halácsy	—	—	—	—	38 Cod	56d Cod	—	—	—	= " ssp. villosulum
panjaoënse Wend.	—	—	—	49 Scor	—	—	—	—	—	
papillare Boiss.	—	—	—	—	—	—	20 Mol* Pal 3 Mol	—	—	
paradoxum (M.Bieb.) G.Don	—	207 Bris	178 Mol	91 Bris	—	37 Bris	—	—	—	
parciflorum Viv.	—	—	—	—	—	52 Scor	—	—	—	

pardoi Loscos ex Willk.	—	—	—	—	—	82 All	7 All*	—	8	*=ampeloprasum Eur perhaps var. of ampeloprasum
parnassicum (Boiss.) Halácsy	—	—	—	—	—	64 Cod	—	—	—	
" ssp. minoicum Zahar.	—	—	—	—	—	—	Cod	—	—	=dentiferum (BPS 1991)
parvulum Vved.	—	132 Minu	120 Hap	—	—	—	—	—	—	
paterfamilias Boiss.	—	—	170 Porr	—	—	—	—	—	41a	USSR=waldsteinii BM=rotundum ssp. rotundum
pauciflorum Viv. ex Gren. et Godron	—	—	—	—	—	—	52 Scor	—	—	=parciflorum (Stearn 1978)
pauli Vved.	—	203 Brevic	+ Mol	—	—	—	—	Brevic	—	
pedemontanumWilld.	—	—	—	—	—	15 Rhiz	Nark	—	—	=narcissiflorum; RMF=Nark
pekinense Prokh.	94 Porr	—	—	—	—	—	—	—	—	=sativum
pendulinum Ten.	—	—	—	—	—	36 Bris	16 Bris*	—	—	*=triquetrum
pentadactyli Brullo, Pavone et Spampinato	—	—	—	—	—	—	Brev	—	—	IK (Italy) 1989
permixtum Guss.	—	—	—	—	—	30 Mol	25 Mol*	—	—	Eur may=subhirsutum; *=roseum
peroninianum Aznav.	—	—	—	—	23 Brev	—	—	—	—	
pervestitum Klokov	—	—	—	—	—	89 All	—	—	32	
petraeum Kar. et Kir.	—	57 Orei	68 Rhiz	—	—	—	—	Orei	—	
pevtzovii Prokh.	—	—	—	—	—	—	—	Orei	—	IK (1930) Kashgar
phalereum Heldr. et Sart. [-lar-]	—	—	—	—	53 Cod	65 Cod	—	—	—	=staticiforme
phanerantherum Boiss. et Hausskn.	—	—	—	69 All	101 All	—	Pal 18 All	—	60	
" ssp. deciduum Kollm. et Koyuncu	—	—	—	—	101 All	—	—	—	—	
" " phanerantherum	—	—	—	—	101 All	—	—	—	—	
phariense Rendle	55 Rhiz	—	—	—	—	—	—	Orei	—	
philistaeum Boiss.	—	—	—	—	—	—	17 Mol* Pal 4 Mol	—	—	=erdelii
phrygium Boiss.	—	—	—	—	57 Cod	—	—	—	—	
phthioticum Boiss. et Heldr.	—	—	—	—	—	26 Mol	—	—	—	
pictistamineum O.Schwarz	—	—	—	—	66 Cod	—	14 Cod*	—	—	*=paniculatum
pilosum Sibth. et Sm.	—	—	—	—	52 Cod	66 Cod	—	—	—	
pisidicum Boiss. et Heldr.	—	—	—	—	20 Brev	—	—	—	—	=cupani ssp. hirtovaginatum
platakisii Tzan. et Kypriotakis	—	—	—	—	—	—	Cod/Scor	—	—	IK (Crete) 1993
platyphyllum (Diels) Wang et Tang	10a Brom	—	—	—	—	—	—	—	—	=wallichii var. platyphyllum
platyspathum Schrenk	53 Rhiz	44 Orei	54 Rhiz	—	—	—	—	Orei	—	
" ssp. amblyophyllum (Kar. et Kir.) Friesen	—	44 Orei	—	—	—	—	Rhiz	Orei	—	IK (USSR) 1987=amblyophyllum; Asia & NWF=good ssp.

Column 1 Species of Allium & authorities (incl. synonyms)	2 Fl. China	3 Middle Asia	4 Fl. USSR (+= Fl. As. med.)	5 Fl. Iran	6 Fl. Turkey	7 Fl. Europ.	8 Others	9 Fritsch & Friesen	10 B. Math-ew	11 Accepted names for synonyms
platyspathum var. falcatum Regel	54 Rhiz	—	—	—	—	—	—	—	—	=carolinianum
platystemon Kar. et Kir.	—	205 Porph	180 Mol	92 Porph	19 Porph	—	—	Porph	—	=oreophilum
platystylum Regel	—	—	—	—	—	—	—	Orei	—	=platyspathum
plurifoliatum Rendle	41 Rhiz	—	—	—	—	—	—	Ret	—	
" var. stenodon (Nakai et Kitag.) J.M.Xu	41a Rhiz	—	—	—	—	—	—	Ret	—	NWF=plurifoliatum
" " zhegushanse J.M.Xu	41b Rhiz	—	—	—	—	—	—	—	—	
podolicum Blocki ex Racib. et Szafer	—	—	—	—	—	62 Cod	—	—	—	See also Stearn 1978
pogonotepalum Wend.	—	—	—	6 Rhiz	—	—	—	Ret	—	
polyanthum Schult. et Schult.f.	—	—	—	—	—	77 All	7 All*	—	10	*=ampeloprasum
" var. aestivalis (J.J.Rodr.) Pau	—	—	—	—	—	—	—	—	2	=commutatum
polyastrum Diels	—	—	—	—	—	—	?	—	—	IK (China) 1912
" var. platyphyllum Diels	10a Brom	—	—	—	—	—	—	—	—	=wallichii var. platyphyllum
polycormium Lovrić	—	—	—	—	—	—	—	Rhiz	—	Croatia; NWF=invalid name
polyphyllum Kar. et Kir.	54 Rhiz	45 Orei	55 Rhiz	11 Rhiz	—	—	—	Orei	—	=carolinianum (exc. USSR)
polyrhizum Turcz. ex Regel [polyrrh-]	27 Rhiz	42 Caes	47 Rhiz	—	—	—	—	Caes	—	
" var. alabasicum D.S.Wen et S.Chen	—	—	—	—	—	—	—	—	—	=alabasicum, q.v.
ponticum Miscz. ex Grossh.	—	—	165 Porr	—	94 All	—	—	—	38	
popovii Vved.	—	105 Scor	137 Hap	—	—	—	—	—	—	
porphyroprasum Heldr.	—	—	—	—	89 All	—	Pal 17 All	—	41a	BM=rotundum ssp. rotundum; Turk & Pal=scorodoprasum ssp. "
porrum L. (cult.)	93 Porr	—	176 Porr	—	77 All	76 All	7 All* Pal s.n. All	—	1a	*=ampeloprasum
" ssp. bimetrale (Gand.) Breistr.	—	—	—	—	—	—	—	—	2	=commutatum
" " eu-ampeloprasum (Hayek) Breistr.	—	—	—	—	—	76 All	—	—	1	=ampeloprasum
" var. ampeloprasum (L.) Mirbel	—	—	—	—	—	—	—	—	1	=ampeloprasum
potaninii Regel	26 Rhiz	—	—	—	—	—	—	But	—	=ramosum
praemixtum Vved.	—	81 Cep	+ Phyll	—	—	—	Phyll	Cep	—	IK (Central Asia) 1946
praescissum Reichb.	—	135 Cod	105 Hap	—	—	56 Cod	—	—	—	USSR & Eur=paniculatum

prattii C.H.Wright apud. Forb. et Hemsl.	6 Ang	—	—	—	—	—	—	Ang	—	
" var. latifolium Wang et Tang	2 Ang	—	—	—	—	—	—	—	—	=ovalifolium
preslianum Roem. et Schult.	—	—	—	—	—	? 87a All	—	—	41a	Eur=?scorodoprasum ssp. rotundum; BM=?rotundum ssp. rotundum
procerum Trautv.	—	175 Comp	217 Mol	—	—	—	—	Comp	—	=giganteum
prokhanovii (Worosch.) V.Yu.Barkalov	—	—	—	—	—	—	Rhiz	Ret	—	IK (USSR) 1987
× proliferum (Moench) Schrad.	—	—	—	—	—	—	—	Cep	—	=cepa × fistulosum
× proliferum Schrad. ex Willd.										see cepa
proponticum Stearn et Özhatay	—	—	—	—	96 All	91 All	—	—	64	
" var. parviflorum Kollm.	—	—	—	—	96 All	—	—	—	64b	
" " proponticum	—	—	—	—	96 All	—	—	—	64a	
prostratum Trevir.	48 Rhiz	—	41 Rhiz	—	—	—	—	Rhiz	—	
" auct. non Trevir.	—	—	—	—	—	—	Rhiz	—	—	=burjaticum (Friesen 1988)
protensum Wend.	—	193 Kal	—	129 Kal	—	—	—	Kal	—	
pruinatum Link ex Spreng.	—	—	—	—	—	93 All	—	—	69	
pruinosum Cand.	—	—	—	—	? 79 All	—	—	—	2	probably=commutatum
przewalskianum Regel	18 Rhiz	—	—	—	—	—	—	Ret	—	
pseudoalbidum Friesen et Özhatay	—	—	—	—	—	—	—	Rhiz	—	IK (Turkey) 1998
pseudoampeloprasum Miscz. ex Grossh.	—	—	172 Porr	—	82 All	—	—	—	4	
pseudocalyptratum Mouterde	—	—	—	—	—	—	—	—	39	Lebanon; Mt. Hermon
pseudocepa Schrenk	—	77 Cep	89 Cep	—	—	—	—	—	—	=galanthum
pseudocyaneum Grüning	84 Hap	—	—	—	—	—	—	Sacc	—	=thunbergii
pseudoflavum Vved.	—	—	101 Hap	88 Cod	56 Cod	74 Cod	—	—	—	Iran & Eur=stamineum
pseudoglobosum M.Pop. ex Gamajun. [ex Pavl. et Polj.]	—	62 Orei	+ Rhiz	—	—	—	—	Orei	—	=kurssanovii
pseudojaponicum Makino	84 Hap	—	—	—	—	—	Ohwi	Sacc	—	=thunbergii
pseudo-ochroleucum Schur	—	—	—	—	—	—	—	Orei		=ochroleucum
pseudophanerantherum Rech.f.	—	—	—	—	—	—	—	—	21	Syria
pseudopulchellum Omelcz.	—	—	—	—	—	?68 Cod	Cod	—	—	=paczoskianum (Kud 1992); Eur=?flavum
pseudostamineum Kollm. et A.Shmida	—	—	—	—	—	—	?Cod	—	—	IK (Lebanon, Syria) 1977
pseudostrictum Albov	—	—	17 Rhiz	—	—	—	—	Ret	—	USSR=szovitsii; NWF=good sp.
pseudotenuissimum Skv.	46 Rhiz	—	—	—	—	—	—	Ten	—	=tenuissimum

Column 1 Species of Allium & authorities (incl. synonyms)	2 Fl. China	3 Middle Asia	4 Fl. USSR (+= Fl. As. med.)	5 Fl. Iran	6 Fl. Turkey	7 Fl. Europ.	8 Others	9 Fritsch & Friesen	10 B. Math-ew	11 Accepted names for synonyms
pseudoxiphopetalum Wend.	—	31 Camp	—	9 Rhiz	—	—	—	Camp	—	=dolichostylum
pseudozeravschanicum M.Pop. et Vved. [-seraw-]	—	145 Meg	215 Mol	125 Meg	—	—	—	Meg	—	=sarawschanicum (exc. USSR)
pskemense B.Fedtsch.	—	78 Cep	90 Cep	—	—	—	—	Cep	—	
pugetii Gand.	—	—	—	—	—	—	14 Cod*	—	—	=paniculatum
pulchellum auct. ross.	—	—	—	—	55 Cod	68b Cod	—	—	—	=flavum ssp. tauricum (Turk= " " " var. tauricum)
" ssp. guicciardii (Heldr.) Nyman	—	—	—	—	—	68a Cod	—	—	—	=flavum ssp. flavum
" G.Don	—	—	102 Hap	—	58 Cod	71b Cod	—	—	—	Turk & Eur=carinatum ssp. pulchellum
" " var. calabrum N.Terracc.	—	—	—	—	—	—	Cod	—	—	see calabrum
pulchrum Clarke	—	—	—	—	—	—	25 Mol*	—	—	=roseum
pumilum Vved.	—	—	46 Rhiz	—	—	—	—	Ret	—	
purpurascens Losa	—	—	—	—	—	18 Schoe	—	Schoe	—	=schoenoprasum
purpureum Loscos	—	—	—	—	—	90 All	—	—	62a	Eur=variant of sphaerocephalon BM=s. ssp. sphaerocephalon
purshii G.Don	—	—	—	—	—	—	—	—	85	=vineale
pusillum Cirillo ex Ten.	—	—	—	—	—	—	51 Scor	—	—	=obtusiflorum (Stearn 1978)
pustulosum Boiss. et Hausskn.	—	—	—	—	86 All	—	—	—	28	
pyrenaicum Costa et Vayreda	—	—	—	—	—	86 All	7 All*	—	12	* =ampeloprasum
pyrrhorrhizum Airy-Shaw [pyrrhorhizum]	42 Rhiz	—	—	—	—	—	Col	—	—	=mairei; RMF=Col (Hanelt et al. 1992)
qaradaghense Feinbr. [-gense]	—	—	—	77 All	86 All	—	—	—	29	Turk perhaps=pustulosum
qasyunense Mout. [gas-]	—	—	—	—	—	—	Pal 5 Mol	—	—	
raddeanum Regel	72 Schoe	—	82 Rhiz	—	—	—	—	Schoe	—	=schoenoprasum
ramazanicum Parsa	—	—	—	?	—	—	—	Acan	—	Iran poor specimen
ramosum L.	26 Rhiz	—	—	—	—	—	—	But	—	
rechingeri Wend.	—	—	—	36 Scor	—	—	—	—	—	
reconditum J.Pastor, B.Valdes et Munoz	—	—	—	—	—	—	Scor	—	—	IK (Spain) 1983
reflexum Boiss. et Reut.	—	—	—	111 Mel	125 Mel	—	—	Mel	—	=chrysantherum (RMF=?chrys.)
regelianum Beck. ex Iljin.	—	—	160 Porr	—	—	94 All	—	—	68	
regelii Trautv.	—	182 Reg	224 Mol	139 Reg	—	—	—	Reg	—	see Kh & Fr 1994

registanicum Wend.	—	—	—	32 Scor	—	—	—	—	—	
regnieri Maire	—	—	—	—	—	—	4a All*	—	62a	=sphaerocephalum ssp. sphaerocephalum
renardii Regel [renarii, corr. by author]	—	118 Caer	130 Hap	—	—	—	—	—	—	=caesium
reticulatum Siehe	—	—	—	—	21 Brev	—	—	—	—	=callidictyon
" J. et C.Presl	—	—	—	—	—	—	11 Rhiz	—	—	=lineare (Stearn 1978)
reuterianum Boiss.	—	—	—	—	99 All	—	—	—	73	
" var. longicaule O.Schwarz	—	—	—	—	—	—	—	—	71	=stylosum
rhabdotum Stearn	—	—	—	—	—	—	?Schoe	Cep	—	IK (Bhutan) 1960
rhetoreanum Nábelek	—	—	—	—	139 Mel	—	—	Kal	—	
rhodanthum Vved.	—	191 Kal	+Mol	—	—	—	—	Kal	—	Asia =caspium, RMF=caspium ssp. baissunense
rhodiacum Brullo, Pavone et Salmeri	—	—	—	—	—	—	Scor	—	—	IK (Greece) 1992
rhodopeum Velen.	—	—	—	—	38 Cod	56d Cod	—	—	—	=paniculatum ssp. villosulum
rhynchogynum Diels	42 Rhiz	—	—	—	—	—	Col	—	—	=mairei; RMF=Col (Hanelt et al. 1992)
rilaense Panov	—	—	—	—	—	95 All	—	—	80c	BM=?guttatum ssp. dalmaticum Eur=vineale var. capsuliferum
riparium Opiz	—	—	—	—	—	18 Schoe	—	Schoe	—	=schoenoprasum
ritsii Iatrou et Tzan.	—	—	—	—	—	—	Brev	—	—	IK (Greece) 1995
robertianum Kollm.	—	—	—	—	118 All	—	—	—	105	
roborowskianum Regel [-ovsk-]	96 Mol	—	202 Mol	—	—	—	—	Mel	—	China & USSR=decipiens
robustum Kar. et Kir.	—	142 Mel	201 Mol	—	—	—	—	Mel	—	
rollii Terracc.	—	—	—	—	105 All	98 All	—	—	81	Eur=dilatatum Turk & BM=amethystinum
rollovii Grossh.	—	—	—	—	93 All	—	—	—	23	
rosenbachianum Regel	—	143 Meg	212 Mol	118 Meg	—	—	—	Meg	—	see also Fritsch 1993
" Vved.	—	147 Meg	—	—	—	—	—	Meg	—	=rosenorum; see Kh & Fr 1994
" sensu Wend.	—	—	—	—	—	—	—	Meg	—	=jesdianum ssp. angustitepalum
rosenorum R.M.Fritsch	—	147 Meg	—	—	—	—	—	Meg	—	Tajikistan; Kh & Fr 1994
roseum L.	—	—	—	—	15 Mol	23 Mol	25 Mol*	—	—	
" ssp. bulbiferum (DC.) E.F.Warburg	—	—	—	—	—	23 Mol	—	—	—	=roseum
" " persicum Bornm.	—	—	—	Nect	Nect	—	—	—	—	=Nectaroscordum tripedale
" var. bulbiferum DC.	—	—	—	—	15 Mol	23 Mol	—	—	—	=roseum
" " bulbilliferum Vis.	—	—	—	—	15 Mol	23 Mol	25 Mol*	—	—	=roseum

Column 1 Species of Allium & authorities (incl. synonyms)	2 Fl. China	3 Middle Asia	4 Fl. USSR (+= Fl. As. med.)	5 Fl. Iran	6 Fl. Turkey	7 Fl. Europ.	8 Others	9 Fritsch & Friesen	10 B. Mathew	11 Accepted names for synonyms
roseum var. carneum (Targ.-Tozz.). Reichb	—	—	—	—	—	23 Mol	25 Mol*	—	—	=roseum
" " cassium (Boiss.) Regel	—	—	—	—	11 Mol	—	—	—	—	=cassium
" " grandiflorum Briq.	—	—	—	—	15 Mol	—	25 Mol*	—	—	=roseum
" " humile Sommier	—	—	—	—	—	23 Mol	—	—	—	=roseum
" " insulare Gen.	—	—	—	—	—	23 Mol	25 Mol*	—	—	=roseum
" " majale (Ten.) Regel	—	—	—	—	15 Mol	—	—	—	—	=roseum
" " puberulum Regel	—	—	—	—	11 Mol	—	—	—	—	=cassium
" " tourneuxii Boiss.	—	—	—	—	—	—	25 Mol* Pal 7 Mol	—	—	*=roseum
" auct.	—	—	—	—	—	—	18b Mol*	—	—	=subhirsutum ssp. subvillosum
rothii Zucc.	—	—	—	—	—	—	Pal 28 Mel	Mel	—	
rotundum auct.	—	—	—	—	—	—	8 All*	—	—	=dregeanum
" L.	—	—	168 Porr	74 All	89 All	87a All	Pal 17 All	—	41	Eur & Turk=scorodoprasum ssp. rotundum
" ssp. ampeloprasoides Miscz.	—	—	—	—	—	—	—	—	25	=gramineum
" " commutatum (Guss.) Nyman	—	—	—	—	—	81 All	—	—	2	=commutatum
" " gramineum Miscz.	—	—	167 Porr	—	—	—	—	—	—	=erubescens
" " jajlae (Vved.) B.Mathew	—	—	—	—	—	—	—	—	41b	
" " multiflorum Cadevall	—	—	—	—	—	77 All	—	—	—	=polyanthum
" " " Rouy	—	—	—	—	—	—	7 All*	—	—	=ampeloprasum
" " rotundum	—	—	—	—	—	—	—	—	41a	
" " scoroprasoides Miscz.	—	—	168 Porr	—	—	—	—	—	—	=rotundum
" " waldsteinii (G.Don) Soó or Richter	—	—	—	—	89 All	87b All	—	—	41c	Turk & Eur=scorodoprasum ssp. waldsteinii
" var. melleum Miscz.	—	—	—	—	89 All	—	—	—	41b	Turk=scorodoprasum ssp. jajlae; BM=rotundum ssp. jajlae
" " preslianum (Roem. et Schult.) Regel	—	—	—	—	—	—	—	—	41a	=?rotundum ssp. rotundum
" " scaberrimum (Serres) Asch. et Graebn.	—	—	—	—	—	—	—	—	9	=scaberrimum
" " waldsteinianum Schult. et Schult. f.	—	—	—	—	89 All	—	—	—	41c	Turk=scorodoprasum ssp. waldsteinii BM=rotundum ssp. waldsteinii

" " waldsteinii (G.Don) Fiori	—	—	—	—	—	—	—	—	41c	" " " "
" Ledeb. p.p.	—	—	170 Porr	—	—	—	—	—	—	=waldsteinii
rouyi Gaut.	—	—	—	—	—	53 Scor	—	—	—	
roxburghii Kunth	25 Rhiz	—	—	—	—	—	—	But	—	=tuberosum
roylei Stearn	—	—	—	14 Rhiz	—	—	6 Schoe	Cep?	—	Hook. rev. Stearn 1945
rubellum M.Bieb.	—	112 Avu	114 Hap	43 Scor	29 Scor	43 Scor	—	—	—	
" ssp. scabrellum Vved.	—	—	116 Hap	—	—	—	—	—	—	=scabrellum
" var. grandiflorum Boiss.	—	116 Avu	—	48 Scor	—	—	—	—	—	=griffithianum
" " " Ledeb.	—	—	115 Hap	—	—	—	—	—	—	USSR=albanum; FOK=rubellum
" " parviflorum Ledeb. p.p.	—	—	113 Hap	—	—	—	—	—	—	=syntamanthum
" " " p.p.	—	—	114 Hap	—	29 Scor	—	—	—	—	=rubellum
" " stellatum C.Koch	—	—	—	43 Scor	—	—	—	—	—	=rubellum
" Schmalh. [Shmal'g.]	—	—	115 Hap	—	—	—	—	—	—	USSR=albanum; FOK=rubellum
rubens Baker	—	—	—	14 Rhiz	—	—	—	Cep?	—	=roylei
" Schrad. ex Willd.	—	39 Rhiz	42 Rhiz	—	—	2 Rhiz	—	Rhiz	—	
rubicundum Willd. ex Kunth	—	—	—	—	—	—	25 Mol*	—	—	=roseum
rubrovittatum Boiss. et Heldr.	—	—	—	—	—	101 All	—	—	53	
" var. occidentale Rouy ex Willk.	—	—	—	—	—	—	—	—	p. 127	perhaps =pruinatum
rudbaricum Boiss. et Buhse	—	—	167 Porr	73 All	89a All	—	—	—	44	=erubescens (Turk vol. 10)
rude J.M.Xu	66 Rhiz	—	—	—	—	—	—	Orei	—	
rudolfii I.Turakulov	—	167 Acmo	—	—	—	—	?	Acmo	—	IK (Tianshan) 1986; see Kh & Fr 1994
ruhmerianum Asch. ex Dur. et Barratte	—	—	—	—	—	—	26 Mol* Rhyn	—	—	Rhyn see Hanelt et al. 1992
rupestre Steven	—	—	109 Hap	—	43 Cod	59 Cod	—	—	—	
rupestristepposum Friesen	—	—	—	—	—	—	—	Ret	—	IK (Siberia) 1992
rupicola Boiss. ex Mouterde	—	—	—	—	73 Cod	—	—	—	—	
rupicolum Boiss. [sphalm.]	—	—	—	—	73 Cod	—	—	—	—	=rupicola
ruprechtii Boiss.	—	—	71 Rhiz	—	—	—	—	Orei	—	=saxatile
sabulosum Stev. ex Bge. in Goeb. [ex Kar. et Kir.]	—	104 Scor	141 Hap	35 Scor	—	55 Scor	—	—	—	
sabzakense Wend.	—	—	—	40 Scor	—	—	—	—	—	
sacculiferum Maxim.	84 Hap	—	3 Rhiz	—	—	—	—	Sacc	—	China=thunbergii
sairamense Regel	91 Hap	125 Caer	134 Hap	—	—	—	—	—	—	=schoenoprasoides
salesovii Regel	—	—	—	—	—	—	—	Rhiz	—	=tytthocephalum
salinum A.I.Baranov et Skvortsov	—	—	—	—	—	—	Rhiz	Caes	—	IK (China) 1965
salota J.Dostal	—	—	—	—	—	—	?	Cep	—	IK (cult.) 1984; NWF=ascalonicum auct.

Column 1 Species of Allium & authorities (incl. synonyms)	2 Fl. China	3 Middle Asia	4 Fl. USSR (+= Fl. As. med.)	5 Fl. Iran	6 Fl. Turkey	7 Fl. Europ.	8 Others	9 Fritsch & Friesen	10 B. Math-ew	11 Accepted names for synonyms
salsuginis Radić	—	—	—	—	—	—	All	—	—	Radić 1989
salsum Skvortsov et A.I.Baranov	31 Rhiz	—	—	—	—	—	—	Caes	—	China=bidentatum, NWF=salinum
salthynicum Tscholok.	—	—	—	—	—	—	Rhiz	Orei	—	IK (Caucasus) 1965 =gunibicum (Kud 1992)
samurense Tscholok.	—	—	—	—	—	—	Rhiz	Orei	—	IK (Caucasus) 1967
sandrasicum Kollm., Özhatay et Bothmer	—	—	—	—	80 All	—	—	—	17	
sannineum Gombault	—	—	—	—	—	—	—	—	94	Lebanon, Israel
saposhnikovii E.V.Nikitina	—	158 Acmo	—	—	—	—	—	Acmo	—	IK (Tianshan) 1962; see Kh & Fr 1994
sarawschanicum Regel [serav-]	—	145 Meg	214 Mol	125 Meg	—	—	—	Meg	—	see Fritsch 1996
sardoum Moris	—	—	—	—	106 All	97b All	—	—	80b	=guttatum ssp. sardoum
sativum L. (cult.)	94 Porr	—	163 Porr	—	74 All	75 All	6 All* Pal s.n. All	—	19a	
" ssp. ophioscorodon (Link) J.Holub	—	—	—	—	—	75 All	—	—	—	=sativum var. ophioscorodon
" var. ophioscorodon (Link) Döll	—	—	—	—	—	75 All	6b All*	—	—	
" " sativum	—	—	—	—	—	—	6a All*	—	—	
" auct.	—	—	162 Porr	—	—	—	—	—	—	=longicuspis
satoanum Kitag.	48 Rhiz	—	—	—	—	—	—	Rhiz	—	=prostratum
savarinii Sennen	—	—	—	—	—	—	9 All*	—	48	=baeticum
savii Parl.	—	—	—	—	—	—	Cod	—	—	Italy, France; BPSS 1994
savranicum Bess. [ex Oxner]	—	—	71 Rhiz	—	—	4 Rhiz	—	—	—	=saxatile
saxatile M.Bieb. (USSR p.p.)	—	—	71 Rhiz	—	—	4 Rhiz	—	Orei	—	
" " " "	—	—	70 Rhiz	—	—	—	—	—	—	=marschallianum
" auct.	—	—	+ Rhiz	—	—	—	—	—	—	=dshungaricum
saxicola Kitag.	51 Rhiz	—	—	—	—	—	—	Rhiz	—	China=senescens
scaberrimum Serres	—	—	—	—	—	78 All	—	—	9	
scabrellum Boiss. et Buhse	—	113 Avu	116 Hap	44 Scor	—	—	—	—	—	Asia & Iran=umbilicatum
scabriflorum Boiss.	—	—	—	—	113 All	—	5 All*	—	89	*=ascalonicum
scabriscapum Boiss. et Kotschy	—	13 Ret	21 Rhiz	5 Rhiz	1 Rhiz	—	—	Camp	—	
" Vved. p.p.	—	—	+ Rhiz	—	—	—	—	—	—	=eriocoleum

scabrum Gilib.	—	—	104 Hap	—	—	—	—	—	—	=oleraceum
" Gilli	—	—	—	50 Scor	—	—	—	—	—	=gillii
schachimardanicum Vved.	—	161 Acmo	+ Mol	—	—	—	—	Acmo	—	see Kh & Fr 1994
schergianum Boiss.	—	—	—	—	70 Cod	—	—	—	—	
schischkinii K.Sobol.	—	—	—	—	—	—	Rhiz	Ret	—	IK (Altai) 1949; see Friesen 1988
schmitzii Coutinho	—	—	—	—	—	19 Schoe	—	Schoe	—	
" var. duriminium Coutinho	—	—	—	—	—	—	—	Schoe	—	=schoenoprasum, see Friesen 1996
schoenoprasoides Regel	91 Hap	125 Caer	134 Hap	26 Scor	—	—	—	—	—	
schoenoprasum L.	72 Schoe	75 Schoe	82 Rhiz	17 Schoe	5 Schoe	18 Schoe	12 Schoe* Pal s.n. Schoe	Schoe	—	see Friesen 1996
" ssp. riparium (Čelak.) Hayek	—	—	—	—	—	18 Schoe	—	Schoe	—	=schoenoprasum
" " sibiricum (L.) Čelak.	—	—	—	—	—	18 Schoe	—	Schoe	—	Friesen 1996=schoenoprasum
" var. alvarense Hyl.	—	—	—	—	—	18 Schoe	—	Schoe	—	=schoenoprasum
" " buhseanum (Regel) Boiss.	—	—	—	17 Schoe	5 Schoe	—	—	Schoe	—	=schoenoprasum
" " foliosum Regel	—	—	—	—	—	—	Ohwi	Schoe	—	NWF=schoenoprasum
" " orientale Regel	—	—	—	—	—	—	Ohwi	Schoe	—	=maximowiczii
" " scaberrimum Regel	72a Schoe	76 Schoe	+ Rhiz	—	—	—	—	Schoe	—	Asia & +=karelinii
" " shibutsuense Kitam.	—	—	—	—	—	—	Ohwi	Schoe	—	=maximowiczii var. shibutsuense
" " sibiricum Garcke	—	—	—	—	—	18 Schoe	—	Schoe	—	Friesen 1996=schoenoprasum
" " " auct. non Garcke	—	—	—	—	—	—	Schoe	—	—	Friesen 1988; =altyncolicum
" f. scaberrimum Kar. et Kir.	—	76 Schoe	—	—	—	—	—	—	—	=karelinii
schrenkii Regel	23 Rhiz	2 Ret	14 Rhiz	—	—	—	—	Ret	—	China, Asia & USSR=strictum
" auct.	—	—	+ Rhiz	—	—	—	—	—	—	=bogdoicolum & oreoprasoides
schubertii auct. non Zucc.	—	—	+ Mol	—	—	—	—	Kal	—	+=bucharicum, RMF=protensum
" Vved.	—	193 Kal	219 Mol	129 Kal	—	—	—	Kal	—	Asia, Iran and RMF=protensum
" Zucc.	—	—	—	—	141 Kal	—	28 Mel* Pal 29 Kal	Kal	—	
schugnanicum Vved.	—	144 Meg	+ Mol	—	—	—	—	Meg	—	see also Fritsch 1993
scopulicolum Font Quer [-cola]	—	—	—	—	—	76 All	—	—	2	Eur prob.=ampeloprasum BM=commutatum
scorodoprasum L.	—	—	161 Porr	—	89 All	87 All	—	—	40	
" ssp. babingtonii (Borrer) Nyman	—	—	—	—	—	76 All	—	—	—	=ampeloprasum
" " jajlae (Vved.) Stearn	—	—	—	—	89 All	87c All	—	—	41b	BM(p.100)=rotundum ssp. jajlae
" " rotundum (L.) Stearn	—	—	—	—	89 All	87a All	Pal 17 All	—	41a	BM=rotundum ssp. rotundum
" " scorodoprasum	—	—	—	—	89 All	87d All	—	—	—	
" " waldsteinii (G.Don) Stearn	—	—	—	—	89 All	87b All	—	—	41c	BM= " " waldsteinii

Column 1 Species of Allium & authorities (incl. synonyms)	2 Fl. China	3 Middle Asia	4 Fl. USSR (+= Fl. As. med.)	5 Fl. Iran	6 Fl. Turkey	7 Fl. Europ.	8 Others	9 Fritsch & Friesen	10 B. Math-ew	11 Accepted names for synonyms
scorodoprasum var. babingtonii (Borrer) Regel	—	—	—	—	—	—	76 All	—	—	=ampeloprasum (Stearn 1978)
" auct.	—	—	—	—	—	—	6b All*	—	—	=sativum var. ophioscorodon
scorzonerifolium Desf. ex DC. [-nerae-]	—	—	—	—	—	34 Mol	—	—	—	
" var. xericiense (Pérez Lara) R.Fernandes	—	—	—	—	—	34 Mol	—	—	—	
scotostemon Wend.	—	—	—	115 Meg	—	—	—	Thaum	—	
scrobiculatum Vved.	—	108 Scor	139 Hap	—	—	—	—	—	—	
scythicum Zoz	—	—	—	—	—	94 All	—	—	?68	perhaps=regelianum
segetum Jan ex Schult. et Schult.f.	—	—	—	—	105 All	96 All	2 All*	—	81	=amethystinum (*=guttatum)
seirotrichum Ducell. et Maire	—	—	—	—	—	—	13 Cod*	—	—	=trichocnemis
semenovi[i] Regel	70 Rhiz	69 Ann	80 Rhiz	—	—	—	—	Ann	—	
" Wend.	—	—	—	16 Schoe	—	—	—	—	—	=fedschenkoanum
semiretschenskianum Regel	88 Hap	121 Caer	127 Hap	—	—	—	—	—	—	=pallasii
senescens L.	51 Rhiz	36 Rhiz	44 Rhiz	—	—	—	—	Rhiz	—	
" ssp. glaucum (Schrad. ex Poir.) Dostal	—	—	—	—	—	—	—	Rhiz	—	NWF=lusitanicum or senescens ssp. montanum (Europe)
" " glaucum (Schrad.) Friesen	—	—	—	—	—	—	—	Rhiz	—	NWF=senescens s.s. (Siberia, Mongolia)
" " montanum (Fries) Holub	—	—	—	—	—	3 Rhiz	—	Rhiz	—	see Stearn 1978; NWF=lusitanicum
" var. brevipedunculatum Regel	—	—	43 Rhiz	—	—	—	—	—	—	=tytthocephalum
" " glaucum (Schrad.) Regel	—	—	—	—	—	—	Rhiz	—	—	NWF=senescens s.str.
seravschanicum Regel										see sarawschanicum
serbicum Vis. et Pančić	—	—	—	—	41 Cod	—	—	—	—	=pallens ssp. pallens
sergii Vved.	—	202 Brevic	183 Mol	—	—	—	—	Brevic	—	see Kh & Fr 1994
serotinum Lapeyr.	—	—	—	—	—	—	3 Rhiz	Rhiz	—	NWF=lusitanicum; Stearn 1978 =senescens ssp. montanum
" Don. ex Hook.f.	—	—	—	—	—	—	—	—	—	=Aneilema scapiflorum Wight (Commelin.); IK (India) 1892
" Zahar.	—	—	—	—	—	—	70 Cod	—	—	=tardans (Stearn 1978)
serrulatum Boiss. in sched.	—	—	—	—	—	—	—	—	—	In IK=atroviolaceum
" f. gracilior Boiss. et Buhse	—	—	—	—	—	—	—	—	44	=erubescens

setaceum Waldst. et Kit.	—	—	—	—	30 Scor	—	—	—	—	=moschatum
setifolium Schrenk	59 Rhiz	49 Orei	59 Rhiz	—	—	—	—	Orei	—	
severtzovii Vved.	—	168 Acmo	—	—	—	—	—	Acmo	—	=severtzovioides
severtzovioides R.M.Fritsch	—	168 Acmo	—	—	—	—	—	Acmo	—	Uzbekistan; see Kh & Fr 1994
sewerzowii Regel [severtzovii]	—	156 Acmo	203 Mol	—	—	—	—	Acmo	—	see Kh & Fr 1994
shatakiense Rech.f.	—	—	—	—	138 Mel	—	—	Mel	—	
shelkovnikovii Grossh.	—	—	—	99 Acan	—	—	—	Acan	—	Iran may=akaka or haemanthoides
sibiricum L.	72 Schoe	—	82 Rhiz	—	5 Schoe	—	—	Schoe	—	=schoenoprasum; see Friesen 1996
sibthorpianum Schult. et Schult.f.	—	—	—	—	49 Cod	—	—	—	—	
siculum Ucria	—	—	—	—	Nect	Nect	—	—	—	=Nectaroscordum siculum
sieberianum Schult. et Schult.f.	—	—	—	—	12 Mol	—	23 Mol*	—	—	=neapolitanum
sieheanum Hausskn. ex Kollm.	—	—	—	—	35 Scor	—	—	—	—	
sikkimense Baker	35 Rhiz	—	—	—	—	—	—	Ret	—	
simethis Lévl.	12 Brom	—	—	—	—	—	—	—	—	=macranthum
simile Regel	97 Mol	160 Acmo	198 Mol	—	—	—	—	Acmo	—	China & USSR=fetis[s]owii; see Kh & Fr 1994; imperfectly known
sinaiticum Boiss.	—	—	—	—	—	—	Pal 22 All	—	98	
sindjarense Boiss. et Hausskn. ex Regel	—	—	—	33 Scor	s.n. Cod	—	Pal 12 Cod	—	—	Turk imperfectly known
singulifolium Rech.f.	—	—	—	—	s.n. Mol	—	—	Mini	—	Turk imperfectly known; RMF=?karataviense
sinkiangense Wang et Tang	95 Mol	—	—	—	—	—	—	Mel	—	
sintenisii Freyn	—	—	—	—	110 All	—	—	—	43	
siphonanthum J.M.Xu	19 Rhiz	—	—	—	—	—	—	Ret	—	
sipyleum Boiss.	—	—	—	—	48 Cod	58 Cod	—	—	—	
sivasicum Özhatay et Kollm.	—	—	—	—	36 Scor	—	—	—	—	
smithii Nyman	—	—	—	—	49 Cod	—	—	—	—	=sibthorpianum
songpanicum J.M.Xu	83 Hap	—	—	—	—	—	—	—	—	
sordidiflorum Vved.	—	22 Camp	+ Rhiz	—	—	—	—	Camp	—	
sosnovskyanum Miscz. ex Grossh. [sosnowskyanum]	—	—	—	—	119 All	—	—	—	114	
spathaceum Steudel ex A.Richard	—	—	—	—	—	—	18c Mol*	—	—	=subhirsutum ssp. spathaceum
spathulatum Khassanov et R.M.Fritsch	—	—	—	—	—	—	All	—	—	(Kyrgyzstan) Fritsch et al. 1998
speciosum Cirillo	—	—	—	—	—	—	29b Mel*	Mel	—	RMF=nigrum,*=n. ssp. nigrum
sphaerocephaloides Foucaud ex Sav.	—	—	—	—	—	—	—	—	62a	=sphaerocephalon ssp. sphaerocephalon

Column 1 Species of Allium & authorities (incl. synonyms)	2 Fl. China	3 Middle Asia	4 Fl. USSR (+= Fl. As. med.)	5 Fl. Iran	6 Fl. Turkey	7 Fl. Europ.	8 Others	9 Fritsch & Friesen	10 B. Mathew	11 Accepted names for synonyms
sphaerocephalon L. [-lum]	—	—	159 Porr	—	95 All	90 All	4 All*	—	62	
" ssp. arvense (Guss.) Arcang.	—	—	—	—	95 All	90b All	—	—	62c	
" " curtum (Boiss. et Gaill.) Duyfjes	—	—	—	—	97 All	—	4c All*	—	67	Turk & BM=curtum
" " durandoi (Batt. et Trab.) Duyfjes	—	—	—	—	—	—	4b All*	—	62d	
" " ebusitanum (Font Quer) Rosselló, Mus et Torres	—	—	—	—	—	—	All	—	—	IK (Spain) 1993
" " eusphaerocephalum Briq.	—	—	—	—	—	—	4a All*	—	62a	=sphaerocephalon ssp. sphaerocephalon
" " rollii (Terracc.) K.Richter	—	—	—	—	—	—	—	—	81	=amethystinum
" " rumelicum Formánek	—	—	—	—	—	—	—	—	80a	=?guttatum ssp. guttatum
" " sardoum (Moris) K.Richter	—	—	—	—	—	—	—	—	80b	=guttatum ssp. sardoum
" " sphaerocephalon	—	—	—	—	95 All	90a All	4a All*	—	62a	
" " trachypus (Boiss. et Spruner) K.Richter or Stearn	—	—	—	—	95 All	90c All	—	—	62b	
" " viridi-album (Tineo) K.Richter	—	—	—	—	—	—	—	—	62c	=sphaerocephalon ssp. arvense
" var. arvense (Guss.) Gren. et Godron	—	—	—	—	—	—	—	—	62c	=sphaerocephalon ssp. arvense
" " bulbiferum Willk.	—	—	—	—	—	—	—	—	62a	= " ssp. sphaerocephalon
" " bulbilliferum Loret et Barrandon	—	—	—	—	—	—	—	—	62a	= " " "
" " descendens (L.) Regel	—	—	—	—	95 All	—	—	—	62a	= " " "
" " durandoi Batt. et Trab.	—	—	—	—	—	—	4b All*	—	62d	= " " durandoi
" " genuinum Coutinho	—	—	—	—	—	—	—	—	62a	= " " sphaerocephalon
" " sardoum (Moris) Regel or Fiori	—	—	—	—	106 All	97b All	—	—	80b	=guttatum ssp. sardoum
" " trachypus (Boiss. et Spruner) Boiss.	—	—	—	—	95 All	90c All	—	—	62b	=sphaerocephalon ssp. trachypus
" " typicum Regel	—	—	—	—	95 All	—	4a All*	—	62a	= " " sphaerocephalon
" " viridi-album auct. non Tineo	—	—	—	—	—	—	4c All*	—	67b	BM=curtum ssp. palaestinum; *=sphaerocephalon ssp. curtum

" " viridi-album Tineo	—	—	—	—	95 All	90b All	4a All*	—	62c	= " " arvense (*=s. ssp. sphaerocephalon)
sphaeropodum [-don] Klokov	—	—	—	—	—	68b Cod	—	—	—	perhaps=flavum ssp. tauricum
spirale Willd.	51 Rhiz	—	44 Rhiz	—	—	—	—	Rhiz	—	China & USSR=senescens
spirophyllum Wend.	—	—	—	55 Scor	—	—	—	—	—	
splendens Willd. ex Schult.f. [ex Roem. et Schult.]	—	—	12 Rhiz	—	—	—	Ohwi	Ret	—	
" Miq.	—	—	—	—	—	—	Ohwi	—	—	=chinense
" ssp. prokhanovii Worosch.	—	—	—	—	—	—	Rhiz	Ret	—	=prokhanovii (BM)
sprengeri Regel	—	—	—	—	—	—	—	Rhiz?	—	NWF=insufficiently known (not in Pal)
spurium G.Don	51 Rhiz	—	44 Rhiz	—	—	—	—	Rhiz	—	=senescens
stamineum Boiss.	—	—	99 Hap	88 Cod	64 Cod	74 Cod	14 Cod* Pal 13 Cod	—	—	*=paniculatum
" ssp. decaisnei (C.Presl) Kollm.	—	—	—	—	—	—	Pal 13a Cod	—	—	=decaisnei
" " stamineum	—	—	—	—	—	—	Pal 13 Cod	—	—	
" var. alpinum Post	—	—	—	—	73 Cod	—	—	—	—	=?rupicola
" " hymettium (Boiss.) Boiss.	—	—	—	—	—	—	73 Cod	—	—	=hymettium (Stearn 1978)
" " nigropedunculatum Opphr.	—	—	—	—	41 Cod	—	14 Cod* Pal 10 Cod	—	—	Pal=pallens; *=paniculatum Turk=pallens ssp. pallens
staticiforme Sibth. et Sm.	—	—	—	—	53 Cod	65 Cod	—	—	—	
stearnianum Koyuncu,Özhatay et Kollm.	—	—	—	—	87 All	—	—	—	87	
" ssp. stearnianum	—	—	—	—	87 All	—	—	—	87a	
" " vanense Kollm. et Koyuncu	—	—	—	—	87 All	—	—	—	87b	
stearnii Pastor et Valdés	—	—	—	—	—	—	Cod	—	—	IK (Spain, Portugal) 1983
stellerianum Willd. [stelleranum]	—	—	40 Rhiz	—	—	7 Rhiz	—	Rhiz	—	
" ssp. tuvinicum Friesen	—	—	—	—	—	—	Rhiz	Rhiz	—	=tuvinicum (Friesen 1988)
" Ledeb. var. beta	—	—	42 Rhiz	—	—	—	—	Rhiz	—	= rubens
stenodon Nakai et Kitag.	41a Rhiz	—	—	—	—	—	—	Ret	—	China=plurifoliatum var. stenodon; NWF=?cyaneum
stenopetalum Boiss. et Kotschy	—	—	—	—	126 Mel	—	—	Mel	—	
" var. pumilum Parsa	—	—	—	101 Acan	—	—	—	—	—	=minutiflorum
stenophyllum Schrenk	—	74 Schoe	85 Rhiz	—	—	—	—	Schoe	—	=oliganthum
stephanophorum Vved.	—	33 Camp	33 Rhiz	—	—	—	—	Camp	—	
stevenii var. alpha Ledeb.	—	—	72 Rhiz	—	—	—	—	Orei	—	=globosum
" " beta "	—	—	71 Rhiz	—	—	—	—	Orei	—	=saxatile
" " " "	—	60 Orei	+ Rhiz	—	—	—	—	Orei	—	=dshungaricum

Column 1 Species of Allium & authorities (incl. synonyms)	2 Fl. China	3 Middle Asia	4 Fl. USSR (+= Fl. As. med.)	5 Fl. Iran	6 Fl. Turkey	7 Fl. Europ.	8 Others	9 Fritsch & Friesen	10 B. Mathew	11 Accepted names for synonyms
stevenii var. gamma Ledeb.	—	—	70 Rhiz	—	—	—	—	Orei	—	=marschallianum
" " " "	—	—	+ Rhiz	—	—	—	—	Orei	—	=petraeum
" " delta "	—	—	68 Rhiz	—	—	—	—	Orei	—	=petraeum
" " eta "	—	—	67 Rhiz	—	—	—	—	Orei	—	=condensatum
stipitatum Regel	—	148 Meg	207 Mol	121 Meg		—	—	Meg	—	see Fritsch 1996
stocksianum Boiss.	—	—	—	57 Scor	—	—	—	—	—	
" var. persicum Boiss.	—	100 Mult	—	59 All	—	—	—	—	100	=borszczowii
stojanovii Kovatschev	—	—	—	—	—	96 All	—	—	81	=amethystinum
stoliczkii Regel	18 Rhiz	—	—	—	—	—	—	Ret	—	China=przewalskianum
stracheyi Baker	—	—	—	—	—	—	8 Rhiz	Orei	—	Hook. rev. Stearn 1945
stramineum Boiss. et Reuter	—	—	—	—	—	34 Mol	s.n. Mol*	—	—	Eur=scorzonerifolium
straussii Bornm.	—	—	—	113 Mel	129 Mel	—	—	Mel	—	=colchicifolium
strictum Schrader	23 Rhiz	2 Ret	14 Rhiz	—	—	11 Rhiz	—	Ret	—	Eur=lineare
" Ledeb.	—	—	17 Rhiz	—	—	—	—	Ret	—	=szovitsii
" var. anodon Boiss.	—	—	17 Rhiz	—	2 Rhiz	—	—	Ret	—	=szovitsii
" " brachyodon Boiss.	—	—	18 Rhiz	—	—	—	—	Ret	—	=brachyodon
stupposum Bornm.	—	—	—	—	14 Mol	—	—	—	—	=longisepalum var. laceratum
stylosum O.Schwarz	—	—	—	—	98 All	—	—	—	71	
suaveolens Jacq.	—	—	—	—	—	10 Rhiz	—	Orei	—	
" var. ericetorum (Thore) Coutinho	—	—	—	—	—	—	8 Rhiz	Orei	—	=ericetorum (Stearn 1978)
" " ochroleucum (Waldst. et Kit.) Fiori	—	—	—	—	—	—	8 Rhiz	Orei	—	=ericetorum (Stearn 1978)
subalbidum Jordan et Fourreau	—	—	—	—	—	—	25 Mol*	—	—	=roseum
subangulatum Regel	27 Rhiz	—	—	—	—	—	—	Caes	—	China=polyrhizum
subhirsutum L.	—	—	—	—	8 Mol	30 Mol	18,18a Mol*	—	—	*=subhirsutum ssp. subhirsutum
" ssp. album auct.	—	—	—	—	—	—	18b Mol*	—	—	*= " ssp. subvillosum
" " ciliare Maire et Weiller	—	—	—	—	—	—	18a Mol*	—	—	*= " ssp. subhirsutum
" " graecum Richter	—	—	—	—	—	—	18a Mol*	—	—	*= " ssp. subhirsutum
" " obtusitepalum (Svent.) Kunkel	—	—	—	—	—	—	Mol	—	—	IK (Canary Is.) 1971
" " permixtum Richter	—	—	—	—	—	—	25 Mol*	—	—	=roseum

" " spathaceum (Steud. ex A.Richard) Duyfjes	—	—	—	—	—	—	18c Mol*	—	—	
" " subvillosum (Salzmann ex Schult. et Schult.f.) Duyfjes	—	—	—	—	—	—	18b Mol*	—	—	
" " trifoliatum (Cirillo) Arcang.	—	—	—	—	9 Mol	—	—	—	—	=trifoliatum
" " " Asch. et Graeb.	—	—	—	—	—	—	18a Mol*	—	—	=subhirsutum ssp. subhirsutum
" var. album auct.	—	—	—	—	—	—	18b Mol*	—	—	=subhirsutum ssp. subvillosum
" " barcense Maire et Weiller	—	—	—	—	—	—	19 Mol*	—	—	=longanum
" " ciliatum Briquet	—	—	—	—	—	—	18a Mol*	—	—	=subhirsutum ssp. subhirsutum
" " glabrum Regel	—	—	—	—	—	—	23 Mol*	—	—	=neapolitanum
" " graecum [(d'Urv.)] Regel	—	—	—	—	9 Mol	—	18a Mol*	—	—	Turk=trifoliatum; * =subhirsutum ssp. subhirsutum
" " hellenicum Hausskn.	—	—	—	—	—	—	18a Mol*	—	—	*= " " "
" " hirsutum Regel	—	—	—	—	—	—	18a Mol*	—	—	*= " " "
							Pal 2 Mol			Pal =trifoliatum var. hirsutum
" " spathaceum Regel	—	—	—	—	—	—	18c Mol*	—	—	=subhirsutum ssp. spathaceum
" " subvillosum Ball	—	—	—	—	—	—	18b Mol*	—	—	= " ssp. subvillosum
" " " (Salzm. ex Schult. et Schult.f.) Batt. et Trab.	—	—	—	—	—	32 Mol	—	—	—	=subvillosum
" " trifoliatum[(Cirillo)] Batt. et Trab.	—	—	—	—	9 Mol	—	18a Mol*	—	—	Turk=trifoliatum; *=subhirsutum ssp. subhirsutum
" " typicum Regel	—	—	—	—	—	—	18a Mol*	—	—	*= " " "
" " vernale Bonnet et Barratte	—	—	—	—	—	—	18b Mol*	—	—	*= " " subvillosum; =subvillosum in Stearn 1978
" auct.	—	—	—	—	9 Mol	—	18b Mol*	—	—	Turk =trifoliatum *=subhirsutum ssp. subvillosum
subnotabile Wend.	—	—	—	76 All	—	—	—	—	35	
subquinqueflorum Boiss. p.p.	—	—	108 ?Hap	—	43 Cod	—	—	—	—	USSR=?kunthianum; Turk=rupestre
" " p.p.	—	—	—	—	50 Cod	—	—	—	—	=tauricolum
subtilissimum Ledeb.	60 Rhiz	50 Orei	60 Rhiz	—	—	—	—	Orei	—	
subvillosum Salzm. ex Schult. et Schult.f.	—	—	—	—	—	32 Mol	18b Mol*	—	—	*=subhirsutum ssp. subvillosum
subvineale Wend.	—	—	—	71 All	104 All	—	—	—	86	Turk=vineale
sulcatum Delile in Redouté	—	—	—	—	—	—	23 Mol*	—	—	=neapolitanum
" DC.	—	—	—	—	12 Mol	—	—	—	—	= "
sulphureum Vved.	—	15 Ret	+ Rhiz	—	—	—	—	Ret	—	
sulvia Buch.-Ham. ex D.Don	25 Rhiz	—	—	—	—	—	—	But	—	=tuberosum
supranisianum F.S.Sailer	—	—	—	—	—	—	?	—	—	IK Supp.18 (Austria) 1841

Column 1 Species of Allium & authorities (incl. synonyms)	2 Fl. China	3 Middle Asia	4 Fl. USSR (+= Fl. As. med.)	5 Fl. Iran	6 Fl. Turkey	7 Fl. Europ.	8 Others	9 Fritsch & Friesen	10 B. Math-ew	11 Accepted names for synonyms
suvorovii Nábelek	—	—	—	119 Meg	—	—	—	—	—	=jesdianum
suworowii [suvorovii] Regel	—	169 Acmo	204 Mol	116 Meg	—	—	—	Acmo	—	see Kh & Fr 1994
svetlanae Vved. ex Z.N.Filimonova	—	126 Caer	—	—	—	—	?	—	—	IK (Pamir-Alai) 1982
synnotii G.Don [-tia in IK, mistake]	—	—	—	—	—	—	—	—	47	see BM under dregeanum
syntamanthum C.Koch [syntha-]	—	—	113 Hap	42 Scor	—	—	—	—	—	
szechuanicum Wang et Tang	36 Rhiz	—	—	—	—	—	—	Ret	—	China=cyaneum; NWF=?cyaneum
szovitsii Regel	—	—	17 Rhiz	—	2 Rhiz	—	—	Ret	—	
" auct. non Regel	—	—	—	—	—	—	Rhiz	—	—	=pseudostrictum (Kud 1992)
szurulense Lerchenf.	—	—	—	—	—	—	—	—	—	=ochroleucum (Fl. Român.)
taciturnum Vved.	—	140 Vved	+ Hap	—	—	—	—	—	—	
taeniopetalum M.Pop. et Vved.	—	172 Acmo	211 Mol	—	—	—	—	Acmo	—	see Kh & Fr 1994
" ssp. mogoltavicum (Vved.) R.M.Fritsch et Khassanov	—	—	—	—	—	—	—	Acmo	—	Fritsch et al. 1998
" ssp. turakulovii R.M.Fritsch et Khassanov	—	—	—	—	—	—	—	Acmo	—	(Kyrgyzstan) Fritsch et al. 1998
taishanense J.M.Xu	49 Rhiz	—	—	—	—	—	—	Rhiz	—	
talassicum Regel	—	59 Orei	69 Rhiz	—	—	—	—	Orei	—	
talijevii Klokov	—	—	—	—	—	83 All	—	—	7	
talyschense Miscz. ex Grossh.	—	—	171 Porr	79 All	—	—	—	—	24	
tanguticum Regel	85 Hap	—	—	—	—	—	—	—	—	
taquetii Lévl. et Vaniot	84 Hap	—	—	—	—	—	—	Sacc	—	=thunbergii
tardans Greuter et Zahar.	—	—	—	—	—	70 Cod	—	—	—	
tardiflorum Kollm. et A.Shmida	—	—	—	—	—	—	?Cod	—	—	IK (Israel) 1990
tartaricum										see tataricum
tashkenticum Khassanov & R.M.Fritsch	—	157 Acmo	—	—	—	—	—	Acmo	—	Uzbekistan; Kh & Fr 1994
tataricum Boiss.	—	—	26 Rhiz	—	—	—	—	—	—	=xiphopetalum
" Ledeb.	—	—	32 Rhiz	—	—	—	—	—	—	=inderiense
" L.f. [tartaricum]	26 Rhiz	—	36 Rhiz	—	—	—	—	But	—	USSR=odorum; China & NWF=ramosum
" auct.	—	—	27 Rhiz	—	—	—	—	—	—	=barszczewskii
" " p.p.	—	—	34 Rhiz	—	—	—	—	—	—	=tenuicaule
" " p.p.	—	—	24 Rhiz	—	—	—	—	—	—	=inconspicuum

" " p.p.	—	—	33 Rhiz	—	—	—	—	—	—	=stephanophorum
" "	—	—	—	—	—	—	14 Cod*	—	—	=paniculatum
" "	—	—	—	—	—	—	—	Camp	—	=inderiense
" " var. longiradiatum Regel	—	28 Camp	29 Rhiz	—	—	—	—	—	—	=longiradiatum
tauricolum Boiss. [-cola]	—	—	—	—	50 Cod	—	—	—	—	
tauricum (Besser ex Reichb.) Grossh.	—	—	—	—	55 Cod	68b Cod	—	—	—	=flavum ssp. tauricum
tchefouense O.Deb.	47 Rhiz	—	—	—	—	—	—	Ten	—	=anisopodium
tchihatschewii Boiss.	—	—	—	—	33 Scor	—	—	—	—	
tchongchanense Lévl.	10 Brom	—	—	—	—	—	Rhiz	—	—	China=wallichii; Airy-Shaw 1931= bulleyanum var. tchongchanense
tekesicolum Regel	—	—	—	—	—	—	—	Camp	—	=teretifolium
tel-avivense Eig	—	—	—	—	—	—	30 Mel* Pal 26 Mel	Mel	—	*=orientale; RMF=good sp.
tenorii Spreng.	—	—	—	—	—	—	23 Mol	—	—	=roseum (Stearn 1978)
tenue G.Don	88 Hap	—	127 Hap	—	—	—	—	—	—	=pallasii
" Regel	—	—	121 Hap	—	—	—	—	—	—	=griffithianum
tenuicaule Regel	—	35 Camp	34 Rhiz	10 Rhiz	—	—	—	Camp	—	
tenuiflorum Ten.	—	—	—	—	—	57b Cod	14 Cod*	—	—	Eur=pallens ssp. tenuiflorum *=paniculatum
" var. pseudotenuiflorum Pamp.	—	—	—	—	—	—	14 Cod*	—	—	*= "
tenuifolium Pohl	—	—	—	—	—	—	—	Schoe	—	NWF=schoenoprasum
tenuissimum L.	—	—	50 Rhiz	—	—	—	—	Ten	—	
" Turcz.	—	—	48 Rhiz	—	—	—	—	—	—	=bidentatum
teretifolium Regel	14 Rhiz	65 Orei	74 Rhiz	—	—	—	—	Camp	—	
thessalicum Brullo, Pavone, Salmeri et Tzanoudakis	—	—	—	—	—	—	Scor	—	—	BPST 1994; Greece
thomsonii Baker	54 Rhiz	45 Orei	—	11 Rhiz	—	—	—	Orei	—	=carolinianum
thracicum Halácsy et Gheorghieff	—	—	—	—	—	69 Cod	—	—	—	=melanantherum
thricocephalum Nábelek										see trichocephalum
thunbergii G.Don	84 Hap	—	—	—	—	—	Ohwi	Sacc	—	
" Regel	—	—	226 Cal	—	—	—	—	—	—	=neriniflorum [nerinifolium]
tianschanicum Rupr.	—	55 Orei	66 Rhiz	—	—	—	—	Orei	—	
tibeticum Rendle	35 Rhiz	—	—	—	—	—	—	Ret	—	=sikkimense
tmoleum O.Schwarz	—	—	—	—	89 All	—	—	—	—	perhaps=scorodoprasum ssp. waldsteinii
togashii Hara [togasii]	—	—	—	—	—	—	Ohwi	Rhiz	—	

Column 1 Species of Allium & authorities (incl. synonyms)	2 Fl. China	3 Middle Asia	4 Fl. USSR (+= Fl. As. med.)	5 Fl. Iran	6 Fl. Turkey	7 Fl. Europ.	8 Others	9 Fritsch & Friesen	10 B. Math-ew	11 Accepted names for synonyms
tokaliense Kamelin et Levichev	—	165 Acmo	—	—	—	—	—	Acmo	—	=motor; see Kh & Fr 1994
tortifolium Batt. et Trab.	—	—	—	—	—	—	7 All*	—	1	=ampeloprasum
tourneuxii Boiss.	—	—	—	—	—	—	25 Mol*	—	—	=roseum
" Chabert	—	—	—	—	—	—	21 Mol*	—	—	
trachyanthum Griseb.	—	—	—	—	s.n. All	—	—	—	—	Turk=imperfectly known; see BM under 54 junceum
trachycoleum Wend.	—	—	—	78 All	88 All	—	Pal 16a All	—	26	
trachypus Boiss. et Spruner	—	—	—	—	95 All	90c All	—	—	62b	=sphaerocephalon ssp. trachypus
trachyscordum Vved. [trachyos-] author's correction	—	16 Ret	22 Rhiz	—	—	—	—	Ret	—	
transtaganum Welw. ex Samp.	—	—	—	—	—	25 Mol	—	—	—	=massaessylum
" " " Rouy	—	—	—	—	—	—	24 Mol*	—	—	=?massaessylum
transvestiens Vved.	—	109 Scor	140 Hap	—	—	—	—	—	—	
trautvetterianum Regel	—	177 Comp	218 Mol	—	—	—	—	Comp	—	
trichocephalum Nábelek	—	—	—	87 Cod	69 Cod	—	—	—	—	=myrianthum
trichocnemis J.Gay	—	—	—	—	—	—	13 Cod*	—	—	
trichocoleum Bornm.	—	—	—	—	—	—	Pal 18 Mol	—	—	=carmeli var. roseum
trifoliatum auct. p.p.	—	—	—	—	—	—	18b Mol*	—	—	=subhirsutum ssp. subvillosum
" Cirillo	—	—	—	—	9 Mol	31 Mol	18a Mol*	—	—	*= " " subhirsutum
" ssp. hirsutum (Regel) Kollm.	—	—	—	—	—	—	18a Mol* Pal 2 Mol	—	—	*= " " "
" " obtusitepalum Svent.	—	—	—	—	—	—	Mol	—	—	= " " obtusitepalum (BM)
" var. hirsutum	—	—	—	—	—	—	18a Mol* Pal 2 Mol	—	—	*= " " subhirsutum
" " sterile Kollm.	—	—	—	—	—	—	18a Mol* Pal 2 Mol	—	—	*= " " "
trifurcatum (Wang et Tang) J.M.Xu	—	—	—	—	—	—	Brom	—	—	IK (China) 1991
trilophostemon Bornm.	—	—	—	—	128 Mel	—	—	Pseud	—	=cardiostemon
tripedale Trautv.	—	—	227 Nect	<u>Nect</u>	<u>Nect</u>	—	—	—	—	Turk & Iran =Nectaroscordum tripedale
tripterum Nasir	—	—	—	—	—	—	Mol	—	—	IK (W.Pakistan) 1975
triquetrum L.	—	—	—	—	17 Bris.	35 Bris.	16 Bris*	—	—	

" ssp. pendulinum (Ten.) Richter	—	—	—	—	—	—	16 Bris*	—	—	=triquetrum
" var. bulbiferum Batt. et Trab.	—	—	—	—	—	—	16 Bris*	—	—	= "
" " pendulinum (Ten.) Regel	—	—	—	—	—	—	16 Bris*	—	—	= "
" " typicum Regel	—	—	—	—	—	—	16 Bris*	—	—	= "
triste Kunth et Bouché	—	—	—	—	—	—	—	Kal	—	sp. unclear
tristissimum Freyn et Sint.	—	—	—	—	43 Cod	—	—	—	—	=rupestre
tristylum Regel	70 Rhiz	69 Ann	—	—	—	—	—	Ann	—	=semenovii
troodi H.Lindb.	—	—	—	—	—	—	Mol	—	—	=cassium var. hirtellum (Cyprus) (BM)
truncatum (Feinbr.) Kollm. et D.Zohary	—	—	—	—	—	—	Pal 16 All	—	18	
tschimganicum O.Fedtsch. p.p.	97 Mol	—	198 Mol	—	—	—	—	Acmo	—	China & USSR=fetis[s]owii
" " p.p. [B.Fedtsch.]	—	166 Acmo	203 Mol	—	—	—	—	Acmo	—	Asia=costatovaginatum, USSR=severtzovii; see OF 1906
tschonoskianum Regel	—	—	—	—	—	—	Rhiz	—	—	IK (Japan) 1875
tschulaktavicum Bajtenov et Nelina	—	—	—	—	—	—	Rhiz	Orei	—	IK (Xinjiang) 1994
tschulpias Regel	—	116 Avu	121 Hap	48 Scor	—	—	—	—	—	=griffithianum
" var. bahri Vved.	—	—	+ Hap	—	—	—	—	—	—	= "
" auct.	—	—	+ Hap	—	—	—	—	—	—	=scabrellum
tsoóngii Wang et Tang	7 Brom	—	—	—	—	—	—	—	—	=hookeri
tubergenii Freyn	—	—	—	—	s.n. Mel	—	—	Mel	—	Turk=imperfectly known
tuberosum Rottl. ex Spreng	25 Rhiz	—	—	—	—	—	Ohwi	But	—	
tubiflorum Rendle	98 Cal	—	—	—	—	—	—	—	—	
tui Wang et Tang	36 Rhiz	—	—	—	—	—	—	Ret	—	=cyaneum, NWF=?cyaneum
tulipifolium Ledeb. [tulipaef-]	96 Mol	141 Mel	202 Mol	—	130 Mel	—	—	Mel	—	China, USSR & Turk=decipiens; (tulipi- Art. 60.8 of Code); RMF=good sp.
tuncelianum (Kollm.) Özhatay, B.Mathew et Siraneci	—	—	—	—	—	—	—	—	20	Turkey
turcicum Özhatay et Cowley	—	—	—	—	—	—	Cod	—	—	IK (Turkey) 1994
turcomanicum Regel [turko-]	—	92 Cost	147 Porr	61 All	—	—	—	—	106	
turkestanicum Regel	—	87 All	142 Hap	65 All	—	—	—	—	115	BM says ?Cod
turtschicum Regel	—	12 Ret	+ Rhiz	—	—	—	—	Ret	—	
tuvinicum (Friesen) Friesen	—	—	—	—	—	—	Rhiz	Rhiz	—	IK (Tuva; Mongolia) 1987
tytthanthum Vved.	—	52 Orei	62 Rhiz	—	—	—	—	Orei	—	

Column 1 Species of Allium & authorities (incl. synonyms)	2 Fl. China	3 Middle Asia	4 Fl. USSR (+= Fl. As. med.)	5 Fl. Iran	6 Fl. Turkey	7 Fl. Europ.	8 Others	9 Fritsch & Friesen	10 B. Mathew	11 Accepted names for synonyms
tytthocephalum Roem. et Schult. [tytho-]	—	—	43 Rhiz	—	—	—	—	Rhiz	—	
ubsicolum Regel	—	—	—	—	—	—	Rhiz	Ret	—	IK (Mongolia) 1887; Friesen 1988
ucrainicum (Kleopow et Oxner) Bordzil.	—	—	—	—	—	39b Oph	—	—	—	=ursinum ssp. ucrainicum
udinicum T.P.Antsupova	—	—	—	—	—	—	Schoe	Schoe	—	IK (Siberia) 1989; NWF=schoenoprasum
ugamii Vved.	—	88 Cost	149 Porr	—	—	—	—	—	107	=filidens
uliginosum Ledeb.	—	—	83 Rhiz	—	—	—	—	Schoe	—	=ledebourianum
" G.Don	25 Rhiz	—	—	—	—	—	—	But	—	=tuberosum
umbilicatum Boiss.	—	113 Avu	—	44 Scor	—	—	—	—	—	
uratense Franch.	89 Hap	—	128 Hap	—	—	—	—	—	—	=macrostemon
urceolatum Regel	—	118 Caer	130 Hap	—	—	—	—	—	—	=caesium
urmiense Kamelin et Seisums	—	—	—	—	—	—	—	Mel	—	IK (Iran) 1996
ursinum L.	—	—	2 Oph	—	—	39 Oph	—	—	—	
" ssp. ucrainicum Kleopow et Oxner	—	—	—	—	—	39b Oph.	—	—	—	
" " ursinum	—	—	—	—	—	39a Oph	—	—	—	
" var. ovalifolium Lacaita	—	—	—	—	—	—	39b Oph	—	—	=u. ssp. ucrainicum (Stearn 1978)
valdecallosum Maire et Weiller	—	—	—	—	—	—	14 Cod*	—	—	=paniculatum
valdesianum Brullo, Pavone et Salmeri	—	—	—	—	—	—	Cod	—	—	IK (Spain) 1996
valentinae Pavl.	—	89 Cost	+Porr	—	—	—	—	—	112	
variegatum Boiss.	—	—	—	—	72 Cod	—	—	—	—	
vavilovii M.Pop. et Vved.	—	79 Cep	+ Phyll 91 Cep	20 Cep	—	—	—	Cep	—	
venustum C.H.Wright	13 Brom	—	—	—	—	—	Cyath	—	—	=cyathophorum; Hanelt & Fr 1994
vernale Tineo	—	—	—	—	—	—	18b Mol*	—	—	=subhirsutum ssp. subvillosum
verrucosum G.Don	—	—	—	—	—	—	—	—	47	see BM under dregeanum
verticillatum Regel	—	206 Vert	185 Mol	—	—	—	—	Vert	—	
veschnjakovii Regel										see weschniakowii
vescum Wend.	—	—	—	89 Cod	—	—	—	—	—	
victorialis L.	1 Ang	—	1 Ang	—	—	22 Ang	—	Ang	—	

" var. angustifolium Hook.f.	6 Ang	—	—	—	—	—	—	—	—	=prattii
" " listera (Stearn) J.M.Xu [listeria]	la Ang	—	—	—	—	—	—	—	—	
" " platyphyllum (Hult.) Makino	—	—	—	—	—	—	Ohwi	—	—	
victoris Vved.	—	184 Reg	+ Mol	—	—	—	—	Reg	—	see Kh & Fr 1994
vineale L.	—	—	153 Porr	—	104 All	95 All	1 All*	—	85	
" ssp. affine (Boiss. et Heldr.) K.Richter	—	—	—	—	—	—	—	—	85	=vineale
" " capsuliferum (Koch) Ceschm. or K.Richter	—	—	—	—	—	95 All	—	—	85	Eur=vineale var. capsuliferum BM=vineale
" " compactum (Thuill.) K.Richter	—	—	—	—	—	—	1 All*	—	85	=vineale
" var. asperiflorum Regel	—	—	—	—	—	—	—	—	85	perhaps=vineale
" " capsuliferum Koch	—	—	—	—	—	95 All	—	—	85	BM=vineale
" " compactum (Thuill.) Cosson [et Germ.]	—	—	—	—	—	95 All	1 All*	—	85	* & BM=vineale
" " nitens (Sauzé et Maillard) Coutinho	—	—	—	—	—	—	—	—	85	=vineale
" " purpureum H.P.G.Koch	—	—	—	—	—	95 All	—	—	85	BM=vineale
" " typicum Asch. et Graebn.	—	—	—	—	—	95 All	1 All*	—	85	=vineale
" " virens Boiss.	—	—	—	—	—	95 All	—	—	85	BM=vineale
vinicolor Wend.	—	—	—	—	—	—	Mel	Mel	—	IK (Iraq) 1973; RMF=?rothii
violaceum Willd.	—	—	—	—	—	—	71a Cod	—	—	=carinatum ssp. carinatum (Stearn 1978)
virens Lam.	—	—	—	—	—	—	63 Cod	—	—	=oleraceum (Stearn 1978)
virescens DC.	—	—	—	—	—	—	63 Cod	—	—	=oleraceum (Stearn 1978)
virgunculae F.Maekawa et Kitam.	—	—	—	—	—	—	Ohwi	Sacc	—	IK (Japan) 1952
viride Grossh.	—	—	151 Porr	68 All	115 All	—	Pal 23 All	—	101	=dictyoprasum (exc. USSR & Iran)
viridiflorum Pobed.	—	206 Vert	+ Mol	—	—	—	—	Vert	—	Asia=verticillatum; RMF=good sp.
viridulum auct.	—	—	125 Hap	—	—	—	—	—	—	=delicatulum
" Ledeb. (USSR p.p.)	—	—	202 Mol	—	—	—	—	Mel	—	USSR=decipiens
viviparum Kar. et Kir.	90 Hap	117 Caer	129 Hap	—	—	—	—	—	—	=caeruleum
vodopjanovae Friesen	—	—	—	—	—	—	Rhiz	Ten	—	IK (Siberia, Mongolia) 1985
volhynicum Bess.	23 Rhiz	—	14 Rhiz	—	—	—	—	—	—	=strictum
vuralii Kit Tan	—	—	—	—	113a All	—	—	—	92	Turk vol. 10
vvedenskyanum Pavl.	—	170 Acmo	+ Mol	—	—	—	—	Acmo	—	see Kh & Fr 1994
vvedenskyi M.Pop	—	—	—	—	—	—	—	Orei	—	=tianschanicum

Column 1 Species of Allium & authorities (incl. synonyms)	2 Fl. China	3 Middle Asia	4 Fl. USSR (+= Fl. As. med.)	5 Fl. Iran	6 Fl. Turkey	7 Fl. Europ.	8 Others	9 Fritsch & Friesen	10 B. Mathew	11 Accepted names for synonyms
wakegi Araki	76 Schoe	—	—	—	—	—	—	Cep	—	China=fistulosum, NWF=proliferum
waldsteinii G.Don	—	—	170 Porr	—	89 All	87b All	—	—	41c	BM=rotundum ssp. waldsteinii Turk & Eur=scorodoprasum ssp. "
wallichianum Steud.	—	—	—	—	—	—	?	—	—	IK (Himalayas); RMF=wallichii
wallichii Kunth	10 Brom	—	—	—	—	—	—	—	—	
" var. platyphyllum (Diels) J.M.Xu	10a Brom	—	—	—	—	—	—	—	—	
walteri Regel	—	192 Kal	193 Mol	103 Acan	—	—	—	Kal	—	Asia & RMF=cristophii, USSR & Iran=bodeanum
warzobicum Kamelin	—	24 Camp	—	—	—	—	Rhiz	Camp	—	IK (Tajikistan) 1980
webbii Clem.	—	—	—	—	55 Cod	68a Cod	—	—	—	=flavum ssp. flavum (Turk=f. ssp. f. var. minus)
weichanicum Palibin	26 Rhiz	—	—	—	—	—	—	But	—	=ramosum
weissii Boiss.	—	—	—	—	53 Cod	65 Cod	—	—	—	=staticiforme
welwitschii Regel	—	—	—	—	—	93 All	—	—	69	=pruinatum
wenchuanense Z.Y.Zhu	—	—	—	—	—	—	Ang	Ang	—	IK (China) 1991
wendelboanum Kollm.	—	—	—	—	34 Scor	—	—	—	—	
wendelboi Matin	—	—	—	—	—	—	—	—	36	Iran
weschniakowii Regel [veschnjakovii]	69 Rhiz	73 Ann	79 Rhiz	—	—	—	—	?Ann	—	
wiedemannianum Regel	—	—	—	—	71 Cod	—	—	—	—	
" var. floribus albidis Regel	—	—	—	—	69 Cod	—	—	—	—	=myrianthum
" " " purpureis Regel	—	—	—	—	71 Cod	—	—	—	—	=wiedemannianum
wildii Heldr.	—	—	—	—	79 All	81 All	—	—	2	=commutatum
willdenowii Kunth [-ovi]	—	—	125 Hap	—	—	—	—	—	—	=delicatulum
willeanum Holmboe	—	—	—	—	—	—	—	—	30	Cyprus
winklerianum Regel	—	187 Reg	222 Mol	136 Reg	—	—	—	Reg	—	
" auct.	—	—	+ Mol	—	—	—	—	—	—	=hissaricum & lipskyanum
woronowii Miscz. ex Grossh.	—	—	—	—	127 Mel	—	—	Acan	—	
xanthicum Griseb. et Schur	—	—	—	—	—	8 Rhiz	—	Orei	—	Eur=ericetorum NWF=ochroleucum
xiangchengense J.M.Xu	—	—	—	—	—	—	?Brom	—	—	IK (China) 1993
xichuanense J.M.Xu	65 Rhiz	—	—	—	—	—	—	Orei	—	

xiphopetalum Aitch. et Baker	—	23 Camp	26 Rhiz	7 Rhiz	—	—	—	Camp	—	
yamarakkyo Honda	—	—	—	—	—	—	—	Sacc	—	NWF=thunbergii
yanchiense J.M.Xu	82 Hap	—	—	—	—	—	—	Orei	—	
yatei Aitch. et Baker [lat-, jat-]	—	182 Reg	224 Mol	139 Reg	—	—	—	Reg	—	=regelii
yesoense Nakai [yezoense]	25 Rhiz	—	—	—	—	—	Ohwi	But	—	=tuberosum
yongdengense J.M.Xu	28 Rhiz	—	—	—	—	—	—	Caes	—	NWF=?salinum
yüanum Wang et Tang	34 Rhiz	—	—	—	—	—	—	Ret	—	
yuchuanii Y.Z.Zhao et J.Y.Chao	—	—	—	—	—	—	Hap	—	—	IK (China) 1989
yunnanense Diels	42 Rhiz	—	—	—	—	—	Col	—	—	=mairei; RMF=Col (Hanelt et al. 1992)
zaprjagajevii Kassacz	—	66 Orei	—	—	—	—	—	Orei	—	IK (USSR) 1973
zebdanense Boiss. et Noë	—	—	—	—	13 Mol	—	—	—	—	
zergericum Khassanov et R.M.Fritsch	—	154 Acmo	—	—	—	—	—	Acmo	—	Kirgizstan; see Kh & Fr 1994
zimmermannianum Gilg	47a Rhiz	—	—	—	—	—	—	Ten	—	NWF=anisopodium, China=a. var. zimmermannianum
Amaryllis caspia Willd.	—	—	220 Mol	—	—	—	—	—	—	=caspium
Caloscordum exsertum Lindl.	81 Hap	—	—	—	—	—	—	—	—	=chinense
C. neriniflorum Herbert	99 Cal	—	—	—	—	—	—	—	—	=neriniflorum
C. nerinifolium Herbert [Callo-]	—	—	226 Cal	—	—	—	—	—	—	=neriniflorum [nerinifolium]
Cepa fistulosa (L.) F.F.Gray	—	—	—	—	—	—	21 Cep	—	—	=fistulosum (Stearn 1978)
C. prolifera Moench	—	—	—	—	—	—	20 Cep	Cep	—	=cepa (Stearn 1978); NWF=proliferum
Crinum caspium Pall.	—	—	220 Mol	131 Kal	—	—	—	Kal	—	=caspium
Getuonis vineale (L.) Rafin.	—	—	—	—	—	—	—	—	85	=vineale
Moly latifolium S.F.Gray	—	—	—	—	—	—	39a Oph	—	—	=ursinum ssp. ursinum (Stearn 1978)
Nectaroscordum koelzii Wend.	—	—	—	Nect	—	—	—	Pseud	—	=koelzii; IK Suppl. 17
N. siculum Schmalh. (also var. dioscoridis Boiss.)	—	—	228 Nect	—	—	—	—	—	—	=dioscoridis
N. tripedale (Trautv.) Traub	—	—	—	Nect	—	—	—	—	—	=tripedale
Nothoscordum inodorum (Aiton) Nicholson	—	—	—	—	—	—	Mol	—	—	=neapolitanum (Stearn 1986)
Nothoscordum inutile (Makino) Kitam.	—	—	—	—	—	—	—	—	—	NWF=?neriniflorum
Ornithogalum afrum Zucc.	—	—	—	—	—	—	29b Mel*	Mel	—	*=nigrum ssp. nigrum; RMF=nigrum
Porrum arenarium (L.) Reichb.	—	—	—	—	—	—	—	—	85	=vineale

Column 1 Species of Allium & authorities (incl. synonyms)	2 Fl. China	3 Middle Asia	4 Fl. USSR (+= Fl. As. med.)	5 Fl. Iran	6 Fl. Turkey	7 Fl. Europ.	8 Others	9 Fritsch & Friesen	10 B. Math-ew	11 Accepted names for synonyms
Porrum ascalonicum auct. non (L.) Reichb.	—	—	—	—	—	—	10 Schoe*	—	—	=cepa
P. descendens (L.) Reichb.	—	—	—	—	—	—	—	—	62a	=sphaerocephalon ssp. sphaerocephalon
P. sativum Reichb.	—	—	—	—	—	—	6a All*	—	—	=sativum var. sativum
Saturnia cernua Maratti	—	—	—	—	—	—	38 Cham	—	—	=chamaemoly (Stearn 1978)
S. etrusca Jordan et Fourr.	—	—	—	—	—	—	38 Cham	—	—	= " " "
S. littoralis Jordan et Fourr.	—	—	—	—	—	—	38 Cham	—	—	= " " "
S. rubrinervis Jordan et Fourr.	—	—	—	—	—	—	38 Cham	—	—	= " " "
S. viridula Jordan et Fourr.	—	—	—	—	—	—	38 Cham	—	—	= " " "
Scilla paradoxa M.Bieb.	—	—	178 Mol	91 Bris	—	—	—	—	—	=paradoxum

BM = Mathew 1996; also pers. comm. BPS 1991 = Brullo, Pavone & Salmeri. BPSS = Brullo, Pavone, Salmeri & Scrugli. BPST = Brullo, Pavone, Salmeri & Tzanoudakis. BPS 1989 = Brullo, Pavone & Spampinato. Fl. Român. = Flora ... România. Hanelt & Fr = Hanelt & Fritsch. Hook. rev. Stearn = Hooker revised by Stearn. IK = *Index Kewensis*. Kh & Fr = Khassanov & Fritsch. Kud = Kudryashova. NWF = Friesen. OF = O.Fedtschenko. RMF = Fritsch.

Sections of *Allium*

Fl. China	Spp. nos.	Fl. USSR	Spp. nos.	Fl. Iran	Spp. nos.	Fl. Turkey	Spp. nos.	Fl. Europaea	Spp. nos.	Fl. Palaestina	Spp. nos.	Allium in Africa	Spp. nos.
Anguinum G.Don All Chinese exc. *victorialis*	1-6	Anguinum (*victorialis*)	1					Anguinum G.Don ex Koch *victorialis*	22				
Bromatorrhiza Ekberg All Chinese exc. *humile*	7-13			(*humile* in Rhiz.)									
Rhiziridium G.Don	14-70	Rhiziridium	3-85	Rhiziridium	1-15	Rhizirideum G.Don ex Koch	1-4	Rhizirideum	1-17				
Schoenoprasum G.Don	71-76	Phyllodolon (Salisb.) Prokh.	86-88	Schoenoprasum Dumort.	16-17	Schoenoprasum	5-6	Schoenoprasum	18-19	Schoenoprasum (cult.)	s.n.	Schoenoprasum	10-12
Cepa Prokh.	77-79	Cepa	89-93	Cepa	18-20	Cepa (Miller) Prokh.	7	Cepa	20-21	Cepa (cult.)	s.n.		
Molium G.Don (Hanelt notes Chinese spp. put by most authors in Melanocrom.)	95-97	Molium	177-225	Molium	90	Molium G.Don ex Koch	8-16	Molium G.Don	23-34	Molium	1-8	Molium	17-27
				Briseis (Salisb.) Stearn (*paradoxum*)	91	Briseis (*triquetrum*)	17	Briseis (*triquetrum* +2 segregated spp.)	35-37			Briseis (*triquetrum*)	16
						Chamaeprason Herm. (or in Molium) (*chamaemoly*)	18	Chamaeprason (or in Molium) (*chamaemoly*)	38				
		Ophioscordon (Wallr.) Vved. (*ursinum*)	2					Ophioscordon (Wallr.) Bubani (*ursinum*)	39				
				Porphyroprason Ekberg (or in Molium) (*oreophilum*)	92	Porphyroprason (or in Molium) (*oreophilum*)	19						

Fl. China	Spp. nos.	Fl. USSR	Spp. nos.	Fl. Iran	Spp. nos.	Fl. Turkey	Spp. nos.	Fl. Europaea	Spp. nos.	Fl. Palaestina	Spp. nos.	Allium in Africa	Spp. nos.
						Brevispatha Valsecchi emend. Garbari (or in Scorodon)	20-24						
Haplostemon Boiss.(Hanelt says Chinese spp. all in Scorodon)	80-92	Haplostemon Boiss.	94-142	Scorodon C.Koch	21-57	Scorodon	25-37	Scorodon	40-55				
				Codonoprasum (Reichb.) Endl.	83-89	Codonoprasum Reichb.	38-73	Codonoprasum	56-74	Codonoprasum	9-14	Codonoprasum	13-15
Porrum	93-94	Porrum	143-176	Allium	58-82	Allium	74-119	Allium	75-104	Allium	15-23	Allium	1-9
		(spp. in Molium)		Acanthoprason Wend. (or in Molium or Melan.)	93-105	Acanthoprason (or in Molium or Melan.)	120-121						
		(spp. in Molium)		Melanocromm-yum Webb et Berth.	106-113	Melanocromm-yum	122-140	Melanocromm-yum	105-109	Melanocromm-yum	24-28	Melanocromm-yum	28-31
		(spp. in Molium)		Megaloprason Wend.	114-128	(spp. in Melan.)							
		(spp. in Molium)		Kaloprason C.Koch	129-132	Kaloprasum (*schubertii*)	141	Kaloprasum (*caspium*)	110	Kaloprasum (*schubertii*)	29	(*schubertii* in Melan.)	
				Thaumasioprason Wend. (Iranian spp. only)	133-135								
		(spp. in Molium)		Regeloprason Wend.	136-139								
Caloscordum (Herb.) Baker	98-99	Caloscordum [Callo-]	226										
		Nectaroscordum (Lindl.) Gren. et Godr.	227-228	Genus *Nectaroscordum* Lindl.				Genus *Nectaroscordum*					

Sections Oreiprason F.Herm. and Reticulato-bulbosa Kamelin are included in Section Rhizirideum in the above table.

The classification adopted by F.O.Khassanov (1996) for column 3 of the list of names is based on that proposed by Hanelt *et al.* (1992), but with many changes of his own. There are too many sections to include it in the above table, so it is outlined separately here with the species numbers.

Section	Species	Section	Species	Section	Species
Subg. Rhizirideum (G.Don ex C.Koch) Wend.		Brevidentia Khassanov et Iengal.	95-96	Compactoprason R.M.Fritsch	175-179
Reticulato-bulbosa Kamelin	1-18	Crystallina " " "	97	Popovia Khassanov et R.M.Fritsch	180
Campanulata Kamelin	19-35	Multicaulea " " "	98-101	Aroidea Khassanov et R.M.Fritsch	181
Rhizirideum G.Don ex Koch	36-39	Scorodon Koch	102-111	Regeloprason Wend.	182-190
Rhizomatosa Egor	40	Avulsea Khassanov	112-116	Kaloprason C.Koch	191-197
Caespitosoprason Friesen	41-42	Caerulea (Omelcz.) Khassanov	117-128	Acanthoprason Wend.	198-199
Tenuissima (Tzag.) Hanelt	43	Kopetdagia Khassanov	129	Miniprason R.M.Fritsch	200
Oreiprason Herm.	44-67	Minuta Khassanov	130-132	Brevicaule R.M.Fritsch	201-203
Petroprason Herm.	68	Brevispatha Valsecchi	133-134	Acaule R.M.Fritsch	204
Annuloprason Egor.	69-73	Codonoprasum Reichb.	135-137	Porphyroprason Ekberg	205
Schoenoprasum Dumort.	74-76	Vvedenskya Kamelin	138-140	Verticillata Kamelin	206
Cepa (Mill.) Prokh.	77-82				
		Subg. Melanocrommyum (Webb et Berth.) Rouy		**Subg. Amerallium** Traub	
Subg. Allium		Melanocrommyum Webb et Berth.	141-142	Briseis (Salisb.) Stearn	207
Allium	83-87	Megaloprason Wend.	143-151		
Costulatae Khassanov et Iengal.	88-94	Acmopetala R.M.Fritsch	152-174		

Note. R.M. Fritsch's sections of subgenus Melanocrommyum and N.W.Friesen's sections of subgenus Rhizirideum are very similar to those of Khassanov for Middle Asia, with the addition of sections Pseudoprason (Wend.) Perss. et Wend. and Thaumasioprason Wend. in subg. Melanocrommyum and of sections Anguinum G.Don ex Koch, Butomissa (Salisb.) Kamelin and Sacculifera P.Gritz. in subg. Rhizirideum. For other section names see Hanelt *et al.* (1992).

J.Radić's (1990) proposed new subgenera Cepa Radić (type *A.cepa* L.) and Steiptoprason Radić (type *A. incensiodorum* Radić) are regarded as synonyms of subgenus Rhizirideum (G.Don ex C.Koch) Wend. if more species than their types are included. Subgenus Polyprason Radić was typified by *A. moschatum* L. independently of section Scorodon Koch.

Abbreviations for section names

(see 'Sections of Allium' for details)

Acan	Acanthoprason	Meg	Megaloprason
Acaul	Acaule	Mel	Melanocrommyum
Acmo	Acmopetala	Micr	Microscordum
All	Allium	Mini	Miniprason
Ang	Anguinum	Minu	Minuta
Ann	Annuloprason	Mol	Molium
Aroi	Aroidea	Mult	Multicaulea
Avu	Avulsea	Nark	Narkissoprason
Brevic	Brevicaule	Nect	genus Nectaroscordum
Brevid	Brevidentia	Oph	Ophioscordon
Brev(is)	Brevispatha	Orei	Oreiprason
Bris	Briseis	Pet	Petroprason
Brom	Bromatorrhiza	Phyl	Phyllodolon
But	Butomissa	Pop	Popovia
Caer	Caerulea	Porph	Porphyroprason
Caes	Caespitosoprason	Porr	Porrum
Cal	Caloscordum	Pseud	Pseudoprason
Camp	Campanulata	Reg	Regeloprason
Cep	Cepa	Ret	Reticulato-bulbosa
Cham	Chamaeprason	Rhiz	Rhizirideum
Cod	Codonoprason	Rhizo	Rhizomatosa
Col	Coleoblastus	Rhyn	Rhynchocarpum
Comp	Compactoprason	Sacc	Sacculifera
Cost	Costulatae	Schoe	Schoenoprason
Crys	Crystallina	Scor	Scorodon
Cyath	Cyathophora	Ten	Tenuissima
Hap	Haplostemon	Thaum	Thaumasioprason
Kal	Kaloprason [um]	Vert	Verticillata
Kop	Kopetdagia	Vved	Vvedenskya

References

Airy-Shaw, H.K. 1931. Allia praesertim Sinensia nova vel minus cognita. *Notes roy. bot. Gard. Edinb.* **16**: 139.

Brullo, S., Pavone, P. & Salmeri, C. 1991. Cytotaxonomical notes on *Allium dentiferum* Webb & Berthelot, an unknown species of the Mediterranean flora. *Bot. Chron.* **10**: 785–796.

Brullo, S., Pavone, P., Salmeri, C. & Scrugli, A. 1994. Cytotaxonomical notes on *Allium savii* Parl. (Alliaceae), a misappreciated Tyrrhenian element. *Candollea* **49**: 271–279.

Brullo, S., Pavone, P., Salmeri, C. & Tzanoudakis, D. 1994. Cytotaxonomical revision of the *Allium obtusiflorum* group (Alliaceae). *Flora medit.* **4**: 179–190.

Brullo, S., Pavone, P. & Spampinato, G. 1989. *Allium pentadactyli* (Liliaceae), a new species from S.Italy. *Willdenowia* **19**: 115–120.

Ekberg, L. 1972. Studies in the genus *Allium.* VI. Bulb structure in the subgenus *Melanocrommyum. Bot. Notiser* **125**: 93–101.

Fedtschenko, O. 1906. Turkestanskie luki. [Turkestan Alliums.] *Progressivnoe Sadovosdtvo I Ogorodnichestvo,* No. 36: 332.

Flora Republicii Socialiste România. 1996. Vol. XI. Bucurest.

Friesen, N.W. [Frizen, N.V.] 1988. *Alliaceae of Siberia: systematics, karyology, chorology.* Novosibirsk Nauka sibir. Otdel. 184 pp. [Russ.]

Friesen, N. 1996. A taxonomic and chorological revision of the genus *Allium* L. sect. *Schoenoprasum* Dumort. *Candollea* **51**: 461–473.

Friesen, N.W. & Özhatay, N. 1998. New taxa and notes on the genus *Allium* subgenus *Rhizirideum* in Turkey. *Feddes Repert.* **109**: 25–31.

Fritsch, R. 1993. Taxonomic and nomenclatural remarks on *Allium* L. subgen. *Melanocrommyum* (Webb & Berth.) Rouy sect. *Megaloprason* Wendelbo. *Candollea* **48**: 417–430.

Fritsch, R.M. 1996. The Iranian species of *Allium* subg. *Melanocrommyum* sect. *Megaloprason* (Alliaceae). *Nordic J. Bot.* **16**: 9–17.

Fritsch, R.M., Khassanov, F.O. & Friesen, N.W. 1998. New taxa, new combinations, and taxonomic remarks on *Allium* L. from Fergan depression, Middle Asia. *Linzer biol. Beitr.* **30**: 281–292.

Garbari, F., Corsi, G. & Masini, A. 1991. Anatomical investigations in the *Allium cupanii* – *A. hirtovaginatum* complex. *Bot. Chron.* **10**: 805–808.

Hanelt, P. & Fritsch, R. 1992. Notes on some infrageneric taxa in *Allium* L. *Kew Bull.* **49**: 559–564.

Hanelt, P., Schultze-Motel, J., Fritsch, R., Kruse, J., Maass, H.I., Ohle, H. & Pistrick, K. 1992. Infrageneric grouping of *Allium* – the Gatersleben approach. pp. 107–123 in: *The genus Allium – taxonomic problems and genetic resources* (eds. Hanelt, P., Hammer, K. & Knüpffer, H.). Inst. Pflanzengenetik und Kulturpflanzenforschung, Gatersleben.

Hooker, J.D. revised by Stearn, W.T. 1945 [1947]. The Alliums of British India. *Herbertia* **12**: 73–84, 174.

Index Kewensis and supplements 1893–1996. Oxford Univ. Press. (Also on CD-ROM.)

Khassanov, F.O. & Fritsch, R.M. 1994. New taxa in *Allium* L., subg. *Melanocrommyum* (Webb et Berth.) Rouy from Central Asia. *Linzer biol. Beitr.* **26**: 965–990.

Kudryashova, G.L. 1992. The synopsis of the species of the genus *Allium* (Alliaceae) from the Caucasus. *Bot. Zhurn.* **77** (4): 86–88. [Russ.]

Mathew, B. 1996. *A review of Allium section Allium.* Royal Botanic Gardens: Kew.

Miceli, P. & Garbari, F. 1991. *Allium aethusanum* and *A. franciniae* (Alliaceae): comparison between two endemic species from Aegadean Islands (Sicily). *Bot. Chron.* **10**: 797–803.

Radić, J. 1989. Slani luk, *Allium salsuginis,* i drugi samonikli lukovi Podbiokovlja. Claves analyticae ac diagnoses latine. With analytic keys and diagnoses in English. *Acta Biokovica* (Makarska) **5**: 5–84. (Many invalid names.)

Radić, J. 1990. Contribution to the knowledge of reproductive peculiarity of genus *Allium* L. in the area Podbiokovlje. *Razprave IV. razreda SAZU* **31**: 247–269.

Ravenna, P. 1991. *Nothoscordum gracile* and *N. borbonicum. Taxon* **40**: 485–487.

Regel, E. 1875. Alliorum adhuc cognitorum monographia. *Acta Horti Petrop.* **3** (2): 1–256.

Seisums, A. 1998. Identity and typification of *Allium magicum, A. nigrum* and *A. roseum* (Alliaceae). *Taxon* **47**: 711–716.

Soldano, A. 1994. Neglected name priorities in the European flora. *Thaiszia* **4**: 119–123.

Stearn, W.T. 1978. European species of *Allium* and allied genera of Alliaceae: a synonymic enumeration. *Ann. Mus. Goulandris* **4**: 83–198.

Stearn, W.T. 1986. *Nothoscordum gracile,* the correct name of *N. fragrans* and the *N. inodorum* of authors (Alliaceae). *Taxon* **35**: 335–337.

Stearn, W.T. 1992. How many species of *Allium* are known? *Kew Mag.* **9**: 180–182.

List of American *Allium* species names

compiled by

Dale W. McNeal
University of the Pacific, Stockton, CA 95211, USA

The genus *Allium* contains 84 recognized native species in North America, north of Mexico. Several additional species are known from Mexico and Central America. While several names have been proposed in the genus from South America, these all seem to represent species of *Nothoscordum.*

The following list is an attempt to help investigators find the generally accepted names of New World species of *Allium* and, where it has been possible, the accepted names of species of *Nothoscordum* that have been described as members of *Allium.* The table shows the name assigned in nine Floras plus records from *Index Kewensis* (IK), the *Gray Herbarium Index* (GHI) and several other sources. It includes about 205 species names, more than half of which are currently considered to be synonyms. The list differs from that of the Old World in that all of the species, except two, are considered to belong to one subgenus, *Amerallium.* As a result, the numbers or + sign in each column simply indicate the acceptance of a name in the particular Flora. The two exceptions are *Allium schoenoprasum* and *A. tricoccum,* which are generally considered to belong to subgenus *Rhizirideum.* No attempt has been made to assign species to the New World sections described by Traub. That system divides the American species in ways that do not reflect their evolutionary relationships. It has largely been ignored by investigators studying the North American species.

Column 1 lists all of the species names found in the literature examined with authorities. As many subspecific taxa as possible are included, however I make no pretence that this category is exhaustive.

The next nine columns show in which Floras the name appears and its species number in that work if the species are numbered. Unlike the Old World list, the New World list does not indicate synonyms listed in the Floras. Names that are synonymous are listed in column 1 and a source for the publication of the name is indicated in column 11. The Floras in columns 2 to 10 are:

Column 2. *Flora of North America,* vol. 23. In Press. (ed. Flora of North America Editorial Committee). Oxford University Press. *Allium* by D. W. McNeal and T. D. Jacobsen.

Column 3. *The Jepson Manual, Higher Plants of California* (ed. J. C. Hickman) 1993. University of California Press, Berkeley. *Allium* by D. W. McNeal.

Column 4. *Vascular Plants of the Pacific Northwest* vol. 1. C.L. Hitchcock, A. Cronquist, F. M. Ownbey, and J. W. Thompson. 1969. University of Washington Press, Seattle. *Allium* by F. M. Ownbey.

Column 5. *Intermountain Flora, Vascular Plants of the Intermountain West, U.S.A.* vol. 6. A. Cronquist, A.H. Holmgren, N. Holmgren, J. Reveal and P.K. Holmgren. 1977. Columbia University Press, New York. *Allium* by A. Cronquist and F. M. Ownbey.

Column 6. *Arizona Flora.* T.H. Kearney and R.H. Peebles. 1969. University of California Press, Berkeley. *Allium* by F. M. Ownbey.

Column 7. *A Flora of New Mexico* vol. 1. W.C. Martin and C.R. Hutchins. 1980. J. Cramer, Germany.

Column 8. *Manual of the Vascular Plants of Texas.* D.S. Correll and M. C. Johnston. 1970. Texas Research Foundation, Renner.

Column 9. *Flora of the Great Plains* (ed. T. M. Barkley). 1986. University of Kansas Press. *Allium* by S. P. Churchill.

Column 10. *Flora of the Southeastern States.* J.K. Small. 1933. Published by the author, New York (Facsimile reprint issued in 2 vols. in 1972 by Hafner.)

Column 11 gives names recognized in Sereno Watson's 1879 monograph on the Liliaceae of North America and Marcus E. Jones's 1902 treatment of *Allium.* These are historical references that any investigator interested in *Allium* in the Western Hemisphere should consult. In order to make the list of species names as complete as possible, other, older Floras were consulted, these are listed in the references that follow the table. Other names are found outside the range of the relevant manuals (eg. Mexico, and Central America) or were described after publication of the relevant Floras. Sources include *Index Kewensis* (IK), the *Gray Herbarium Index* (GHI) or the original publication of the name. The source is followed, where possible, by = and what is, in my judgement, the correct name of the taxon. In some cases I was unable to determine the correct name of a taxon which appears in no Flora; in those cases I have cited a reference for the name.

Some, perhaps most, of Traub's Mexican species will likely be reduced to synonymy as they are studied. These specimens were in the author's personal herbarium and unavailable for study prior to his death. They have since been deposited at the Missouri Botanic Garden. I have not, to date, seen them.

Allium species and synonyms (American species)

Column 1 Species of Allium & authorities (incl. synonyms)	2 FNA	3 California	4 Pacific N. W.	5 Intermt. West	6 Arizona	7 New Mexico	8 Texas	9 Great Plains	10 S. E. States	11 Accepted names for synonyms
aaseae M.Ownbey	51	—	+	15	—	—	—	—	—	
abramsii (H.Traub) D.McNeal	30	+	—	—	—	—	—	—	—	
acetabulum (Raf.) Shinners	—	—	—	—	—	—	—	—	—	GHI 1957
" var. fraseri (M.Ownbey) Shinners	—	—	—	—	—	—	—	—	—	GHI 1957; =canadense var. fraseri
" var. lavendulare (Bates) Shinners	—	—	—	—	—	—	—	—	—	GHI 1957; =canadense var. lavendulare
acuminatum Hook.	80	+	+	12	5	—	—	—	—	
" var. cuspidatum Fern.	—	—	—	—	—	—	—	—	—	GHI 1894; =acuminatum
" var. gracile A.Wood	—	—	—	—	—	—	—	—	—	Wood 1868; =amplectens
allegheniense J.K.Small	—	—	—	—	—	—	—	—	2	=cernuum
ambiguum M.E.Jones	—	—	—	—	—	—	—	—	—	Jones 1902; =obtusum var. obtusum
amplectens Torr.	82	+	+	13	—	—	—	—	—	
anceps Kell.	55	+	—	18	—	—	—	—	—	
" var. lemmonii	—	—	—	—	—	—	—	—	—	Jepson 1921; =lemmonii
" var. aberrans M.E.Jones	—	—	—	—	—	—	—	—	—	Jones 1902; =tolmiei var. tolmiei
angulosum L.	—	—	—	—	—	—	—	—	—	IK 1753
" var. leucorhizum Nutt.	—	—	—	—	—	—	—	—	—	GHI (Arkansas, US) 1835
anserinum Jeps.	—	—	—	—	—	—	—	—	—	Jepson 1921; =parishii
arenicola Osterh.	—	—	—	—	—	—	—	—	—	IK 1900; =geyeri var. tenerum
arenicola J.K.Small	—	—	—	—	—	—	—	—	6	=canadense var. mobilense
aridum Rydb.	—	—	—	—	—	—	—	—	—	IK 1917; =textile
atrorubens S. Wats.	24	+	—	9	—	—	—	—	—	
" var. atrorubens	24a	+	—	9	—	—	—	—	—	
" var. cristatum (S.Wats.) D.McNeal	24b	+	—	—	—	—	—	—	—	Int.W. & AZ =nevadense var. cristatum
" var. inyonis (M.E.Jones) M. Ownbey et H.Aase	—	—	—	9	—	—	—	—	—	Munz 1959; =var. cristatum
" ssp. inyonis (M.E.Jones) H.Traub	—	—	—	—	—	—	—	—	—	Traub 1972b; =var. cristatum
attenuifolium Kellogg	—	—	—	—	—	—	—	—	—	IK 1863; =amplectens
" var. monospermum Jepson	—	—	—	—	—	—	—	—	—	Jepson 1901; =amplectens
austinae M.E.Jones	—	—	—	—	—	—	—	—	—	Jones 1902; =campanulatum
bidwelliae S.Wats.	—	—	—	—	—	—	—	—	—	Watson 1879; =campanulatum
biflorum Larranaga	—	—	—	—	—	—	—	—	—	GHI (Uruguay) 1923
bigelovii S.Wats.	54	+	—	—	12	10	—	—	—	

bisceptrum S.Wats.	42	+	+	10	—	—	—	—	—	
" var. bisceptrum	—	—	—	10	—	—	—	—	—	=bisceptrum
" var. palmeri (S. Wats.) Cronq.	—	—	—	10	—	—	—	—	—	=bisceptrum
" var. utahense M.E.Jones	—	—	—	—	—	—	—	—	—	Jones 1902; =bisceptrum
bivalve (L.) Kuntze	—	—	—	—	—	—	—	—	—	GHI 1898; =Nothoscordum bivalve
" var. andicola (Kunth) Kuntze	—	—	—	—	—	—	—	—	—	GHI 1898; =N. andicola
" var. bangii Kuntze	—	—	—	—	—	—	—	—	—	GHI 1898
" var. flavescens (Kunth) Kuntze	—	—	—	—	—	—	—	—	—	GHI 1898; =N. flavescens
" var. fragrans (Vent.) Kuntze	—	—	—	—	—	—	—	—	—	=N. gracile Stearn
" var. gaudichaudianum (Kunth) Kuntze	—	—	—	—	—	—	—	—	—	GHI 1898; =N. gaudichaudianum
" var. sellowianum (Kunth) Kuntze	—	—	—	—	—	—	—	—	—	GHI 1898; =N. sellowianum
" var. straitum (Jacq.) Kuntze	—	—	—	—	—	—	—	—	—	GHI 1898
bolanderi S.Wats.	76	+	—	—	—	—	—	—	—	
" var. bolanderi	76a	+	—	—	—	—	—	—	—	
" var. mirabile (L.Henders.) D. McNeal	76b	+	—	—	—	—	—	—	—	
" var. stenanthum Jepson	—	—	—	—	—	—	—	—	—	Jepson 1921; =var. bolanderi
bonariense Kuntze	—	—	—	—	—	—	—	—	—	GHI 1908; =N. bonariense
brandegei S.Wats.	44	—	+	16	—	—	—	—	—	
brevistylum S.Wats.	17	—	+	7	—	—	—	—	—	
breweri S.Wats.	—	—	—	—	—	—	—	—	—	Watson 1879; =falcifolium
bullardii Davidson	—	—	—	—	—	—	—	—	—	IK 1923; =campanulatum
burdickii (Hanes) A.G.Jones	—	—	—	—	—	—	—	—	—	Jones 1979; =tricoccum var. burdickii
burlewii Davidson	63	+	—	—	—	—	—	—	—	
californicum Rose	—	—	—	—	—	—	—	—	—	IK (Baja, CA, Mexico) 1890
campanulatum S.Wats.	40	+	+	11	—	—	—	—	—	
" var. bidwelliae Jepson	—	—	—	—	—	—	—	—	—	Jepson 1921; =campanulatum
canadense L.	3	—	—	—	—	—	1	1	4	
" var. canadense	3a	—	—	—	—	—	1a	1a	—	
" var. ecristatum (M.E.Jones) M. Ownbey	3e	—	—	—	—	—	1d	—	—	
" ssp. ecristatum (M.E.Jones) H. Traub et M.Ownbey	—	—	—	—	—	—	—	—	—	Traub 1967; =var. ecristatum
" var. fraseri M.Ownbey	3f	—	—	—	—	—	1e	1b	—	
" ssp. fraseri (M.Ownbey) H.Traub et M.Ownbey	—	—	—	—	—	—	—	—	—	Traub 1967; =var. fraseri
" var. hyacinthoides (Bush) M. Ownbey	3c	—	—	—	—	—	1c	1c	—	

Column 1 Species of Allium & authorities (incl. synonyms)	2 FNA	3 California	4 Pacific N. W.	5 Intermt. West	6 Arizona	7 New Mexico	8 Texas	9 Great Plains	10 S. E. States	11 Accepted names for synonyms
canadense ssp. hyacinthoides (Bush) H.Traub et M.Ownbey	—	—	—	—	—	—	—	—	—	Traub 1967; =var. hyacinthoides
" var. lavendulare (J.M.Bates) M. Ownbey	3d	—	—	—	—	—	—	1d	—	
" ssp. lavendulare (J.M.Bates) H. Traub et M.Ownbey	—	—	—	—	—	—	—	—	—	Traub 1967; =var. lavendulare
" var. mobilense (Regel) M.Ownbey	3b	—	—	—	—	—	1b	1e	—	
" ssp. mobilense (Regel) H.Traub et M.Ownbey	—	—	—	—	—	—	—	—	—	Traub 1967; =var. mobilense
" var. ovoideum Farwell	—	—	—	—	—	—	—	—	—	GHI 1915; =var. canadense
" var. robustum Farwell	—	—	—	—	—	—	—	—	—	GHI 1915; =var. canadense
canescens Beauverd	—	—	—	—	—	—	—	—	—	GHI (Uruguay 1908); =N. canescens
cascadense M.E.Peck	—	—	—	—	—	—	—	—	—	IK 1936; =crenulatum
cernuum Roth	20	—	+	5	9	—	—	2	1	
" var. neomexicana (Rydb.) Macbr.	—	—	—	—	9	6	—	—	—	=cernuum
" ssp. neomexicanum (Rydb.) H. Traub et M.Ownbey	—	—	—	—	—	—	—	—	—	Traub 1967; =cernuum
" var. obtusum Cockll.	—	—	—	—	9	7	—	—	—	=cernuum
" ssp. obtusum (Cockll.) H.Traub et M.Ownbey	—	—	—	—	—	—	—	—	—	Traub 1967; =cernuum
collinum Dougl. in Howell	—	—	—	—	—	—	—	—	—	IK 1902; =fibrillum
columbianum (M.Ownbey et L. Mingrone) P.Peterson, C.Annable et L.Rieseberg	45	—	—	—	—	—	—	—	—	
concinnum K.Brandegee ex M.E. Jones	—	—	—	—	—	—	—	—	—	Jones 1902; =obtusum var. obtusum
constrictum (M.Ownbey et Mingrone) P.Peterson, C.Annable et L.Rieseberg	46	—	—	—	—	—	—	—	—	
continuum J.K.Small	—	—	—	—	—	—	—	—	—	Small 1903; =canadense var. canadense
coreyi M.E.Jones	8	—	—	—	—	—	5	—	—	
cratericola Eastw.	59	+	—	—	—	—	—	—	—	
crenulatum Wieg.	64	—	+	—	—	—	—	—	—	
crispum E.L.Greene	87	+	—	—	—	—	—	—	—	

cristatum S.Wats.	—	—	—	—	—	—	—	—	—	Watson 1879; =atrorubens var. cristatum
croceum Torr.	—	—	—	—	—	—	—	—	—	IK 1859; =Bloomeria crocea
cusickii S.Wats.	—	—	—	—	—	—	—	—	—	Watson 1879; =tolmiei var. tolmiei
cuspidatum Rydb.	—	—	—	—	—	—	—	—	—	GHI 1917; =acuminatum
cuthbertii J.K.Small	9	—	—	—	—	—	—	—	7	
davisiae M.E.Jones	—	—	—	—	—	—	—	—	—	Jones 1902; =lacunosum var. davisiae
decipiens M.E.Jones	—	—	—	—	—	—	—	—	—	Jones 1902; =atrorubens var. cristatum
denticulatum (H.Traub) D.McNeal	29	+	—	—	—	—	—	—	—	
deserticola Woot. et Stand.	—	—	—	—	—	—	—	—	—	IK 1913; =macropetalum
diabolense (M.Ownbey et H.Traub) D.McNeal	37	+	—	—	—	—	—	—	—	
dichlamydeum E.L.Greene	86	+	—	—	—	—	—	—	—	
dictuon H.St.John	79	—	+	—	—	—	—	—	—	
dictyotum E.L.Greene	—	—	—	—	—	—	—	—	—	GHI 1901
diehlii M.E.Jones	—	—	—	—	—	—	—	—	—	Jones 1902; =brandegei
douglasii Hook.	47	—	+	—	—	—	—	—	—	
" var. columbianum M.Ownbey et L.Mingrone	—	—	+	—	—	—	—	—	—	=columbianum
" var. constrictum L.Mingrone et M. Ownbey	—	—	+	—	—	—	—	—	—	=constrictum
" var. douglasii	—	—	+	—	—	—	—	—	—	=douglasii
" var. nevii (S. Wats.) M.Ownbey et L.Mingrone	—	—	+	—	—	—	—	—	—	=nevii
" var. tolmiei H.Traub	—	—	—	—	—	—	—	—	—	Traub 1945; =tolmiei var. tolmiei
drummondi Regel	7	—	—	—	—	—	4	3	—	
" forma asexuale M.Ownbey	—	—	—	—	—	—	—	—	—	GHI 1951; =drummondi
durangoense H.Traub	—	—	—	—	—	—	—	—	—	Traub (Mexico) 1968
elmendorfii M.Ownbey	38	—	—	—	—	—	10	—	—	
equicaeleste H.St.John	—	—	—	—	—	—	—	—	—	IK 1931; =macrum
eurotophilum I.Wiggins	—	—	—	—	—	—	—	—	—	Wiggins (Baja, CA) 1980
falcifolium Hook. et Arn.	66	+	—	—	—	—	—	—	—	
" var. breweri (S.Wats.) M.E.Jones	—	—	—	—	—	—	—	—	—	Jones 1902; =falcifolium
" var. demissum Jepson	—	—	—	—	—	—	—	—	—	Jones 1902; =siskiyouense
fantasmosense H.Traub	—	—	—	—	—	—	—	—	—	Traub (Mexico) 1968
fibrillum M.E.Jones	43	—	+	—	—	—	—	—	—	
fibrosum Rydb.	—	—	—	—	—	—	—	—	—	IK 1897; =geyeri var. tenerum
fimbriatum S.Wats.	35	+	—	—	—	—	—	—	—	
" var. aboriginum Jepson	—	—	—	—	—	—	—	—	—	Jepson 1921; =var. fimbriatum

Column 1 Species of Allium & authorities (incl. synonyms)	2 FNA	3 California	4 Pacific N. W.	5 Intermt. West	6 Arizona	7 New Mexico	8 Texas	9 Great Plains	10 S. E. States	11 Accepted names for synonyms
fimbriatum var. abramsii M.Ownbey et H. Aase ex H.Traub	—	—	—	—	—	—	—	—	—	Traub 1972a; =abramsii
" var. denticulatum M.Ownbey et H.Aase ex H.Traub	—	—	—	—	—	—	—	—	—	Traub 1972a; =denticulatum
" var. diabolense M.Ownbey et H. Aase ex H.Traub	—	—	—	—	—	—	—	—	—	Traub 1972a; =diabolense
" var. fimbriatum	35a	+	—	—	—	—	—	—	—	
" var. mohavense Jepson	35b	+	—	—	—	—	—	—	—	
" ssp. mohavense (Jepson) H. Traub et M.Ownbey	—	—	—	—	—	—	—	—	—	Traub 1967; =var. mohavense
" var. munzii M.Ownbey et H.Aase ex H.Traub	—	—	—	—	—	—	—	—	—	Traub 1972a; =munzii
" var. parryi (S.Wats.) M.Ownbey et H.Aase	—	—	—	—	—	—	—	—	—	Munz 1959; =parryi
" ssp. parryi (S.Wats.) H.Traub	—	—	—	—	—	—	—	—	—	Traub 1967; =parryi
" var. purdyi (Eastw.) M.Ownbey et H.Traub	35c	+	—	—	—	—	—	—	—	
" ssp. purdyi (Eastw.) H.Traub et M.Ownbey	—	—	—	—	—	—	—	—	—	Traub 1967; =var. purdyi
" var. sharsmithae M.Ownbey et H. Aase ex H.Traub	—	—	—	—	—	—	—	—	—	Traub 1972a; =sharsmithae
fragile A.Nels.	—	—	—	—	—	—	—	—	—	IK 1926; =scilloides
fragrans Vent.	—	—	—	—	—	—	—	—	—	=N. gracile Stearn
fraseri (Ownbey) Shinners	—	—	—	—	—	—	—	—	—	IK 1951; =canadense var. fraseri
funiculosum A.Nels.	—	—	—	—	—	—	—	—	—	IK 1934; =geyeri var. geyeri
geyeri S.Wats.	4	—	+	2	1	3	7	4	—	
" var. geyeri	4a	—	+	—	—	—	—	—	—	
" var. graniferum L.Henders.	—	—	—	—	—	—	—	—	—	GHI 1930; =var. tenerum
" var. tenerum M.E.Jones	4b	—	+	—	—	—	—	—	—	
" ssp. tenerum (M.E.Jones) H. Traub et M.Ownbey	—	—	—	—	—	—	—	—	—	Traub 1967; =var. tenerum
glandulosum Link et Otto	75	—	—	—	7	—	—	—	—	
goodingii M.Ownbey	15	—	—	—	6	14	—	—	—	
grandisceptrum Davidson	—	—	—	—	—	—	—	—	—	IK 1924; =unifolium

grossibulbum Beauverd	—	—	—	—	—	—	—	—	—	GHI (Uruguay) 1908
guatemalense H.Traub	—	—	—	—	—	—	—	—	—	IK (Guatemala) 1974
haematochiton S.Wats.	19	+	—	—	—	—	—	—	—	
helleri J.K.Small	—	—	—	—	—	—	—	—	—	Small 1903; =drummondi
hendersonii Robinson et Seaton	—	—	—	—	—	—	—	—	—	GHI 1893; =douglasii
hickmanii Eastw.	81	+	—	—	—	—	—	—	—	
hintoniorum B.L.Turner	—	—	—	—	—	—	—	—	—	IK (Mexico) 1994
hoffmanii M.Ownbey et H.Traub	62	+	—	—	—	—	—	—	—	
howardii H.Traub	—	—	—	—	—	—	—	—	—	Traub (US, Texas) 1967
howellii Eastw.	27	+	—	—	—	—	—	—	—	
" var. clokeyi M.Ownbey et H.Traub	27c	+	—	—	—	—	—	—	—	
" var. howellii	27a	+	—	—	—	—	—	—	—	
" var. sanbenitense (H.Traub) H. Traub et M.Ownbey	27b	+	—	—	—	—	—	—	—	
" ssp. sanbenitense (H.Traub) M. Ownbey et H.Aase	—	—	—	—	—	—	—	—	—	Traub 1972b; =var. sanbenitense
huntiae H.Traub	—	—	—	—	—	—	—	—	—	Traub (Mexico) 1968
hyacinthoides B.F.Bush	—	—	—	—	—	—	—	—	—	IK 1906; =canadense var. hyacinthoides
hyalinum M.K.Curran	84	+	—	—	—	—	—	—	—	
" var. hickmanii Eastw.	—	—	—	—	—	—	—	—	—	Jepson 1921; =hickmanii
" var. praecox Jepson	—	—	—	—	—	—	—	—	—	Jepson 1921; =praecox
idahoense H.Traub	—	—	—	—	—	—	—	—	—	IK 1947; =tolmiei var. tolmiei
inactum (Jepson) H.Traub	—	—	—	—	—	—	—	—	—	Traub 1972b; =sanbornii var. congdonii
incisum Nels. et Macbr.	—	—	—	—	—	—	—	—	—	IK 1913; =lemmonii
intactum Jepson	—	—	—	—	—	—	—	—	—	Jepson 1921; =sanbornii var. congdonii
inyonis M.E.Jones	—	—	—	—	—	—	—	—	—	Jones 1902; =atrorubens var. cristatum
jepsonii (H.Traub) S.Denison et D. McNeal	28	+	—	—	—	—	—	—	—	
johnstonii M.E.Jones ex Jepson	—	—	—	—	—	—	—	—	—	Jepson 1921; =burlewii
kessleri Davidson	—	—	—	—	—	—	—	—	—	IK 1921; =parryi
kunthii G.Don	77	—	—	—	8	9	12	—	—	
lacunosum S.Wats.	78	+	—	—	—	—	—	—	—	
" var. davisiae (M.E.Jones) D. McNeal et M.Ownbey	78d	+	—	—	—	—	—	—	—	
" var. kernensis D.McNeal et M. Ownbey	78b	+	—	—	—	—	—	—	—	
" var. lacunosum	78a	+	—	—	—	—	—	—	—	
" var. micranthum Eastw.	78c	+	—	—	—	—	—	—	—	

Column 1 Species of Allium & authorities (incl. synonyms)	2 FNA	3 California	4 Pacific N. W.	5 Intermt. West	6 Arizona	7 New Mexico	8 Texas	9 Great Plains	10 S. E. States	11 Accepted names for synonyms
latifolium W.Young	—	—	—	—	—	—	—	—	—	IK (America) 1783; = ?
lavendulare J.M.Bates	—	—	—	—	—	—	—	—	—	IK 1916; =canadense var. lavendulare
" var. fraseri (M.Ownbey) Shinners	—	—	—	—	—	—	—	—	—	GHI 1953; =canadense var. fraseri
lemmonii S.Wats.	57	+	+	19	—	—	—	—	—	
lloydiiflorum	—	—	—	—	—	—	—	—	—	GHI 1908; =N. lloydiiflorum
macrantherum Kuntze	—	—	—	—	—	—	—	—	—	GHI 1898 (Paraguay); =N. macrantherum
macropetalum Rydb.	14	—	—	4	4	4	6	—	—	
macrum S.Wats.	49	—	+	—	—	—	—	—	—	
madidum S.Wats.	39	—	+	—	—	—	—	—	—	
mannii H.Traub	—	—	—	—	—	—	—	—	—	Traub (Mexico) 1968
marvinii Davidson	—	—	—	—	—	—	—	—	—	IK 1921; =haematochiton
melliferum H.Traub	—	—	—	—	—	—	—	—	—	Traub (Mexico) 1968
membranaceum M.Ownbey ex Traub	41	+	—	—	—	—	—	—	—	
mexicanum H.Traub	—	—	—	—	—	—	—	—	—	Traub (Mexico) 1968
michoacanum H.Traub	—	—	—	—	—	—	—	—	—	Traub (Mexico) 1970
microscordion J.K.Small	—	—	—	—	—	—	—	—	5	=canadense var. mobilense
minarum Beauverd	—	—	—	—	—	—	—	—	—	GHI (Uruguay) 1908; =N. minarum
minimum M.E.Jones	—	—	—	—	—	—	—	—	—	Jones 1902; =brandegei
mirabile L.Henders.	—	—	—	—	—	—	—	—	—	IK 1930; =bolanderi var. mirabile
miser Piper	—	—	—	—	—	—	—	—	—	Jones 1902; =punctum
mobilense Regel	—	—	—	—	—	—	—	—	—	IK 1875; =canadense var. mobilense
modocense Jepson	—	—	—	—	—	—	—	—	—	Jepson 1921; =parvum
mohavense (Jepson) I.Tidestrom	—	—	—	—	—	—	—	—	—	IK 1935; =fimbriatum var. mohavense
monospermum Jepson ex E.L.Greene	—	—	—	—	—	—	—	—	—	IK 1894; =amplectens
monticola Davidson	26	+	—	—	—	—	—	—	—	
" var. keckii (Munz) M.Ownbey et H.Aase	—	—	—	—	—	—	—	—	—	Munz 1959; =monticola
" ssp. keckii (Munz) H.Traub et M. Ownbey	—	—	—	—	—	—	—	—	—	Traub 1967; =monticola
montigenum Davidson	—	—	—	—	—	—	—	—	—	IK 1920; =peninsulare
munzii (M.Ownbey et H.Aase ex H. Traub) D.McNeal	34	+	—	—	—	—	—	—	—	
mutabile Michx.	—	—	—	—	—	—	—	—	—	Small 1903; =canadense var. mobilense
neomexicana Rydb.	—	—	—	—	—	—	—	—	—	IK 1899; =cernuum

nevadense S.Wats.	23	+	+	8	10	—	—	—	—	
" var. cristatum (S.Wats.) M. Ownbey	—	—	—	—	10	—	—	—	—	Munz 1959; =atrorubens var. cristatum
" ssp. cristatum (S.Wats.) H.Traub	—	—	—	—	—	—	—	—	—	Traub 1967; =var. cristatum
" var. macropetalum M.Peck	—	—	—	—	—	—	—	—	—	Peck 1945; =nevadense
" var. nevadense	—	—	—	—	10	—	—	—	—	=nevadense
nevii S.Wats.	48	—	+	—	—	—	—	—	—	PNW =douglasii var. nevii
nuttallii S.Wats.	—	—	—	—	—	—	—	—	—	Watson 1879; =drummondi
obtectum L.Henders.	—	—	—	—	—	—	—	—	—	IK 1941; =acuminatum
obtusum Lemmon	53	+	—	—	—	—	—	—	—	
" var. conspicuum W.Mortola et D. McNeal	53b	+	—	—	—	—	—	—	—	
" var. obtusum	53a	+	—	—	—	—	—	—	—	
occidentale A.Gray	—	—	—	—	—	—	—	—	—	IK 1867; =amplectens
ownbeyi H.Traub	—	—	—	—	—	—	—	—	—	Traub (Mexico) 1968
oxyphilum Wherry	—	—	—	—	—	—	—	—	—	IK 1925; =cernuum
palmeri S.Wats.	—	—	—	—	13	11	—	—	—	Watson 1879; =bisceptrum
parishii S.Wats.	25	+	—	—	11	—	—	—	—	
" var. keckii Munz	—	—	—	—	—	—	—	—	—	Munz 1935; =monticola
" var. parishii	—	—	—	—	—	—	—	—	—	=parishii
parryi S.Wats.	33	+	—	—	—	—	—	—	—	
parvum Kell.	58	+	+	22	—	—	—	—	—	
" var. brucae M.E.Jones	—	—	—	—	—	—	—	—	—	Jones 1902; =cratericola
" var. jacintense Munz	—	—	—	—	—	—	—	—	—	Munz 1935; =cratericola
passeyi N.Holmgren et A.Holmgren	12	—	—	—	—	—	—	—	—	
pauciflorum Larranaga	—	—	—	—	—	—	—	—	—	GHI (Uruguay) 1923
peirsonii Jepson	—	—	—	—	—	—	—	—	—	Jepson 1921; =monticola
peninsulare Lemmon ex E.L.Greene	88	+	—	—	—	—	—	—	—	
" var. crispum Jepson	—	+	—	—	—	—	—	—	—	Jepson 1921; =crispum
" var. franciscanum D.McNeal et M. Ownbey	88b	+	—	—	—	—	—	—	—	
" var. peninsulare	88a	+	—	—	—	—	—	—	—	
perdulce S.V.Fraser	6	—	—	—	—	13	—	5	—	
" var. perdulce	6a	—	—	—	—	—	3a	—	—	
" var. sperryi M.Ownbey	6b	—	—	—	—	—	3b	—	—	
" ssp. sperryi (M.Ownbey) H.Traub et M.Ownbey	—	—	—	—	—	—	—	—	—	Traub 1967; =var. sperryi
persimile (M.Ownbey) H.Traub	—	—	—	—	—	—	—	—	—	Traub 1967; =tolmiei var. persimile

Column 1 Species of Allium & authorities (incl. synonyms)	2 FNA	3 California	4 Pacific N. W.	5 Intermt. West	6 Arizona	7 New Mexico	8 Texas	9 Great Plains	10 S. E. States	11 Accepted names for synonyms
pictum H.Moldenke	—	—	—	—	—	—	—	—	—	Jones 1979; =tricoccum var. tricoccum
pikeanum Rydb.	—	—	—	—	—	—	—	—	—	IK 1904; =geyeri var. geyeri
platycaule S.Wats.	65	+	—	17	—	—	—	—	—	
platyphyllum I.Tidestrom	—	—	—	—	—	—	—	—	—	IK 1916; =tolmiei var. tolmiei
pleianthum S.Wats.	—	—	+	—	—	—	—	—	—	Watson 1879; =tolmiei var. tolmiei
" var. particolor M.E.Jones	—	—	—	—	—	—	—	—	—	Jones 1902; =parvum
plummerae S.Wats	11	—	—	—	3	—	—	—	—	
poeppigii Kunth	—	—	—	—	—	—	—	—	—	GHI (Chile ?) 1921; =N. poeppigii
potosiense H.Traub	—	—	—	—	—	—	—	—	—	IK 1969
praecox T.Brandegee	83	+	—	—	—	—	—	—	—	
pseudobulbiferum Davidson	—	—	—	—	—	—	—	—	—	IK 1921; =lacunosum var. davisiae
pueblanum H.Traub	—	—	—	—	—	—	—	—	—	Traub (Mexico) 1968
punctum L.Henders.	56	+	—	21	—	—	—	—	—	
purdyi Eastw.	—	—	—	—	—	—	—	—	—	GHI 1938; =fimbriatum var. purdyi
recurvatum Rydb.	—	—	—	—	—	—	—	—	—	GHI 1900; =cernuum
reticulatum Fraser	—	—	—	—	—	—	—	—	—	IK 1827; =textile
" var. deserticola M.E.Jones	—	—	—	—	—	—	—	—	—	Jones 1902; =macropetalum
" var. ecristatum M.E.Jones	—	—	—	—	—	—	—	—	—	Jones 1935; =canadense var. ecristatum
" var. nuttalii M.E.Jones	—	—	—	—	—	—	—	—	—	Jones 1908; =drummondi
" var. playnum M.E.Jones	—	—	—	—	—	—	—	—	—	Jones 1908; =textile
rhizomatum Woot. et Standl.	74	—	—	—	—	8	11	—	—	
robinsonii L.Henders.	69	—	—	—	—	—	—	—	—	
robustum Eastw.	—	—	—	—	—	—	—	—	—	IK 1938; =howellii var. sanbenitense
roguense M.E.Peck	—	—	—	—	—	—	—	—	—	IK 1936; =bolanderi var. mirabile
rubrum Osterh.	—	—	—	—	2	1	—	—	—	=geyeri var. tenerum
runyoni M.Ownbey	5	—	—	—	—	—	2	—	—	
rydbergii Macbr.	—	—	—	—	—	—	—	—	—	IK 1918; =geyeri var. tenerum
sabuicola Osterh.	—	—	—	—	—	—	—	—	—	GHI 1900; =geyeri var. tenerum
sanbenitense H.Traub	—	—	—	—	—	—	—	—	—	IK 1947; =howellii var. sanbenitense
sanbornii A.Wood	22	+	—	—	—	—	—	—	—	
" var. congdonii Jepson	22b	+	—	—	—	—	—	—	—	
" var. inactum H.Traub	—	—	—	—	—	—	—	—	—	Traub 1972b; =var. congdonii
" var. intactum Jepson	—	—	—	—	—	—	—	—	—	Jepson 1921; =var. congdonii
" ssp. intactum (Jepson) H.Traub	—	—	—	—	—	—	—	—	—	=var. congdonii

" var. jepsonii M.Ownbey et H. Aase ex H.Traub	—	—	—	—	—	—	—	—	—	Traub 1972a; =jepsonii
" var. tuolumnense M.Ownbey et H. Aase ex H.Traub	—	—	—	—	—	—	—	—	—	Traub 1972a; =tuolumnense
" var. sanbornii	22a	+	—	—	—	—	—	—	—	
scabridulum Beauverd	—	—	—	—	—	—	—	—	—	GHI (Uruguay) 1908; =N. scabridulum
scaposum Benth.	—	—	—	—	—	—	—	—	—	Traub 1968; =kunthii
schoenoprasum L.	18	—	+	—	—	—	—	—	—	Sect. Schoenoprasum
" var. laurentianum Fernald	—	—	—	—	—	—	—	—	—	Fernald 1950; =schoenoprasum
" var. sibiricum (L.) Hartm.	—	—	—	—	—	—	—	—	—	Fernald 1950; =schoenoprasum
scilloides Dougl. ex S.Wats.	68	—	+	—	—	—	—	—	—	
scissum Nels. et Macbr.	—	—	—	—	—	—	—	—	—	IK 1918; =lemmonii
serra D.McNeal et M.Ownbey	85	+	—	—	—	—	—	—	—	
serratum S.Wats.	—	—	—	—	—	—	—	—	—	Watson 1879; =amplectens
" var. dichlamydeum Jones	—	—	—	—	—	—	—	—	—	Jones 1902; =dichlamydeum
sessile R.E.Fries	—	—	—	—	—	—	—	—	—	GHI (Argentina); =N. sessile
sharsmithae (M.Ownbey ex H.Traub) D.McNeal	31	+	—	—	—	—	—	—	—	
shevockii D.McNeal	32	+	—	—	—	—	—	—	—	
simillimum L.Henders.	50	—	+	14	—	—	—	—	—	
siskiyouense M.Ownbey ex H.Traub	67	+	—	—	—	—	—	—	—	
speculae M.Ownbey	10	—	—	—	—	—	—	—	—	
stellatum Ker.	21	—	—	—	—	—	9	7	—	
stenanthum Drew	—	—	—	—	—	—	—	—	—	IK 1889; =bolanderi var. bolanderi
stoloniferum T.Jacobsen	—	—	—	—	—	—	—	—	—	IK (Mexico) 1979
subbiflorum Colla	—	—	—	—	—	—	—	—	—	IK (Chile) 1836
subsessile Beauverd	—	—	—	—	—	—	—	—	—	GHI (Uruguay) 1908; =N. subsessile
subteretifolium H.Traub	—	—	—	—	—	—	—	—	—	IK (Mexico) 1968
teleponense H.Traub	—	—	—	—	—	—	—	—	—	Traub (Mexico) 1968
tenellum Davidson	—	—	—	—	—	—	—	—	—	IK 1922; =campanulatum
texanum T.M.Howard	—	—	—	—	—	—	—	—	—	IK (Mexico) 1990
textile Nels. et Macbr.	13	—	+	3	—	—	—	—	—	
tolmiei J.G.Baker	60	—	+	20	—	—	—	—	—	
" var. persimile M.Ownbey	60b	—	+	—	—	—	—	—	—	
" var. platyphyllum (I.Tidestrom) M.Ownbey	—	—	+	—	—	—	—	—	—	=var. tolmiei
" var. tolmiei	60a	+	+	—	—	—	—	—	—	
traubii T.M.Howard	—	—	—	—	—	—	—	—	—	IK (Mexico) 1967

Column 1 Species of Allium & authorities (incl. synonyms)	2 FNA	3 California	4 Pacific N. W.	5 Intermt. West	6 Arizona	7 New Mexico	8 Texas	9 Great Plains	10 S. E. States	11 Accepted names for synonyms
tribracteatum Torr.	52	+	—	—	—	—	—	—	—	
" var. andersonii S.Wats.	—	—	—	—	—	—	—	—	—	Watson 1879; =parvum
" var. diehlii M.E.Jones	—	—	—	—	—	—	—	—	—	Jones 1902; =brandegei
" var. parvum Jepson	—	—	—	—	—	—	—	—	—	Jepson 1921; =parvum
tricoccum Solander	1	—	—	—	—	—	—	9	1	Sect. Anguinum; SE =Validallium tricoccum
" var. burdickii Hanes	1b	—	—	—	—	—	—	—	—	
" var. tricoccum	1a	—	—	—	—	—	—	—	—	
triflorum Larranaga	—	—	—	—	—	—	—	—	—	GHI (Uruguay)
tuolumnense (M.Ownbey et H.Aase ex H.Traub) S.Denison et D.McNeal	36	+	—	—	—	—	—	—	—	
uniflorum Larranaga	—	—	—	—	—	—	—	—	—	GHI (Uruguay) 1923
unifolium Kell.	73	+	+	—	—	—	—	—	—	
" var. lacteum E.L.Greene	—	—	—	—	—	—	—	—	—	GHI 1890; =unifolium
validum S.Wats.	16	+	+	6	—	—	—	—	—	
vancourvense Macoun	—	—	—	—	—	—	—	—	—	IK 1888; =crenulatum
watsonii Howell	—	—	—	—	—	—	—	—	—	IK 1902; =crenulatum
yosemitense Eastw.	61	+	—	—	—	—	—	—	—	
zenobiae Cory	—	—	—	—	—	—	—	—	—	IK 1953; =canadense var. mobilense

References

Fernald, M.L. 1950. *Grays Manual of Botany.* 8th ed. pp. 429–432. American Book Co. New York.

Jepson, W.L. 1901. *Flora of Western Middle California.* Encina Publishing Co., Berkeley, CA. pp. 119–120.

Jepson, W.L. 1921. *A Flora of California.* Assoc. Students Store, Univ. of California, Berkeley, CA. pp. 270–280.

Jones, A.G. 1979. A study of wild leek, and recognition of *Allium burdickii* (Liliaceae). *Syst. Bot.* **4**(1):29–43.

Jones, M.E. 1902. (Treatment of *Allium* in) *Contributions to Western Botany* **10**:1–33, 70–77, 83–86, 17 unnumbered pages of figures.

Jones, M.E. 1908. (New varieties of *Allium* in) New species and notes. *Contributions to Western Botany* **12**:79–80.

Jones, M.E. 1935. (A new variety of *Allium* in) New species and notes. *Contributions to Western Botany* **18**:21.

Munz, P.A. 1935. *Manual of Southern California Botany.* Claremont Colleges, CA. pp. 85–88.

Munz, P.A. 1959. *A California Flora.* University of California Press, Berkeley, CA. pp. 1368–1378.

Peck, M.E. 1945. *Allium nevadense* var. *macropetalum. Leafl. W. Bot.* **4**:177.

Small, J.K. 1903. *Flora of the Southeastern United States.* Published by the author. New York.

Traub, H. 1945. New *Allium* names and a deleted species. *Herbertia* **12**:68–70.

Traub, H. 1967. Amaryllid notes, 1967. *Plant Life* **23**: 110.

Traub, H. 1968. New Guatemalan and Mexican Alliums. *Plant Life* **24**:127–141.

Traub, H. 1972a. *Allium* species and varieties. *Plant Life* **28**:63–64.

Traub, H. 1972b. Amaryllid notes, 1972. *Plant Life* **28**:66–67.

Watson, S. 1879. Contributions to American Botany IX. Revision of the North American Liliaceae. *Proc. Amer. Acad. Arts* **14**:213–288.

Wiggins, I.L. 1980. *Flora of Baja California.* Stanford University Press, Stanford, CA. p. 838.

Wood, A. 1868. A sketch of the natural order Liliaceae as represented in the flora of the states of Oregon and California. *Proc. Acad. Nat. Sci.* **20**: 170.